DNA Methylation Protocols

METHODS IN MOLECULAR BIOLOGY™

John M. Walker, SERIES EDITOR

205. **E. coli Gene Expression Protocols,** edited by *Peter E. Vaillancourt,* 2002
204. **Molecular Cytogenetics:** *Methods and Protocols,* edited by *Yao-Shan Fan,* 2002
203. **In Situ Detection of DNA Damage:** *Methods and Protocols,* edited by *Vladimir V. Didenko,* 2002
202. **Thyroid Hormone Receptors:** *Methods and Protocols,* edited by *Aria Baniahmad,* 2002
201. **Combinatorial Library Methods and Protocols,** edited by *Lisa B. English,* 2002
200. **DNA Methylation Protocols,** edited by *Ken I. Mills and Bernie H, Ramsahoye,* 2002
199. **Liposome Methods and Protocols,** edited by *Subhash C. Basu and Manju Basu,* 2002
198. **Neural Stem Cells:** *Methods and Protocols,* edited by *Tanja Zigova, Juan R. Sanchez-Ramos, and Paul R. Sanberg,* 2002
197. **Mitochondrial DNA:** *Methods and Protocols,* edited by *William C. Copeland,* 2002
196. **Oxidants and Antioxidants:** *Ultrastructural and Molecular Biology Protocols,* edited by *Donald Armstrong,* 2002
195. **Quantitative Trait Loci:** *Methods and Protocols*, edited by *Nicola J. Camp and Angela Cox,* 2002
194. **Post-translational Modification Reactions,** edited by *Christoph Kannicht,* 2002
193. **RT-PCR Protocols,** edited by *Joseph O'Connell,* 2002
192. **PCR Cloning Protocols, 2nd ed.,** edited by *Bing-Yuan Chen and Harry W. Janes,* 2002
191. **Telomeres and Telomerase:** *Methods and Protocols,* edited by *John A. Double and Michael J. Thompson,* 2002
190. **High Throughput Screening:** *Methods and Protocols,* edited by *William P. Janzen,* 2002
189. **GTPase Protocols:** *The RAS Superfamily,* edited by *Edward J. Manser and Thomas Leung,* 2002
188. **Epithelial Cell Culture Protocols,** edited by *Clare Wise,* 2002
187. **PCR Mutation Detection Protocols,** edited by *Bimal D. M. Theophilus and Ralph Rapley,* 2002
186. **Oxidative Stress and Antioxidant Protocols,** edited by *Donald Armstrong,* 2002
185. **Embryonic Stem Cells:** *Methods and Protocols,* edited by *Kursad Turksen,* 2002
184. **Biostatistical Methods,** edited by *Stephen W. Looney,* 2002
183. **Green Fluorescent Protein:** *Applications and Protocols,* edited by *Barry W. Hicks,* 2002
182. **In Vitro Mutagenesis Protocols, 2nd ed.,** edited by *Jeff Braman,* 2002
181. **Genomic Imprinting:** *Methods and Protocols,* edited by *Andrew Ward,* 2002
180. **Transgenesis Techniques, 2nd ed.:** *Principles and Protocols,* edited by *Alan R. Clarke,* 2002
179. **Gene Probes:** *Principles and Protocols,* edited by *Marilena Aquino de Muro and Ralph Rapley,* 2002
178. **Antibody Phage Display:** *Methods and Protocols,* edited by *Philippa M. O'Brien and Robert Aitken,* 2001
177. **Two-Hybrid Systems:** *Methods and Protocols,* edited by *Paul N. MacDonald,* 2001
176. **Steroid Receptor Methods:** *Protocols and Assays,* edited by *Benjamin A. Lieberman,* 2001

175. **Genomics Protocols**, edited by *Michael P. Starkey and Ramnath Elaswarapu,* 2001
174. **Epstein-Barr Virus Protocols,** edited by *Joanna B. Wilson and Gerhard H. W. May,* 2001
173. **Calcium-Binding Protein Protocols, Volume 2:** *Methods and Techniques,* edited by *Hans J. Vogel,* 2001
172. **Calcium-Binding Protein Protocols, Volume 1:** *Reviews and Case Histories,* edited by *Hans J. Vogel,* 2001
171. **Proteoglycan Protocols,** edited by *Renato V. Iozzo,* 2001
170. **DNA Arrays:** *Methods and Protocols,* edited by *Jang B. Rampal,* 2001
169. **Neurotrophin Protocols,** edited by *Robert A. Rush,* 2001
168. **Protein Structure, Stability, and Folding,** edited by *Kenneth P. Murphy,* 2001
167. **DNA Sequencing Protocols,** *Second Edition,* edited by *Colin A. Graham and Alison J. M. Hill,* 2001
166. **Immunotoxin Methods and Protocols**, edited by *Walter A. Hall,* 2001
165. **SV40 Protocols,** edited by *Leda Raptis,* 2001
164. **Kinesin Protocols,** edited by *Isabelle Vernos,* 2001
163. **Capillary Electrophoresis of Nucleic Acids, Volume 2:** *Practical Applications of Capillary Electrophoresis,* edited by *Keith R. Mitchelson and Jing Cheng,* 2001
162. **Capillary Electrophoresis of Nucleic Acids, Volume 1:** *Introduction to the Capillary Electrophoresis of Nucleic Acids,* edited by *Keith R. Mitchelson and Jing Cheng,* 2001
161. **Cytoskeleton Methods and Protocols,** edited by *Ray H. Gavin,* 2001
160. **Nuclease Methods and Protocols,** edited by *Catherine H. Schein,* 2001
159. **Amino Acid Analysis Protocols,** edited by *Catherine Cooper, Nicole Packer, and Keith Williams,* 2001
158. **Gene Knockout Protocols,** edited by *Martin J. Tymms and Ismail Kola,* 2001
157. **Mycotoxin Protocols,** edited by *Mary W. Trucksess and Albert E. Pohland,* 2001
156. **Antigen Processing and Presentation Protocols,** edited by *Joyce C. Solheim,* 2001
155. **Adipose Tissue Protocols,** edited by *Gérard Ailhaud,* 2000
154. **Connexin Methods and Protocols,** edited by *Roberto Bruzzone and Christian Giaume,* 2001
153. **Neuropeptide Y Protocols**, edited by *Ambikaipakan Balasubramaniam,* 2000
152. **DNA Repair Protocols:** *Prokaryotic Systems,* edited by *Patrick Vaughan,* 2000
151. **Matrix Metalloproteinase Protocols,** edited by *Ian M. Clark,* 2001
150. **Complement Methods and Protocols,** edited by *B. Paul Morgan,* 2000
149. **The ELISA Guidebook,** edited by *John R. Crowther,* 2000
148. **DNA–Protein Interactions:** *Principles and Protocols* **(2nd ed.)**, edited by *Tom Moss,* 2001
147. **Affinity Chromatography:** *Methods and Protocols*, edited by *Pascal Bailon, George K. Ehrlich, Wen-Jian Fung, and Wolfgang Berthold,* 2000
146. **Mass Spectrometry of Proteins and Peptides,** edited by *John R. Chapman,* 2000
145. **Bacterial Toxins:** *Methods and Protocols*, edited by *Otto Holst,* 2000

METHODS IN MOLECULAR BIOLOGY™

DNA Methylation Protocols

Edited by

Ken I. Mills, PhD

*Department of Haematology,
University of Wales College of Medicine, Cardiff, UK*

and

Bernard H. Ramsahoye, MD, PhD

*Department of Haematology,
Western General Hospital, Edinburgh, UK*

Humana Press ✴ Totowa, New Jersey

© 2002 Humana Press Inc.
999 Riverview Drive, Suite 208
Totowa, New Jersey 07512
humanapress.com

All rights reserved. No part of this book may be reproduced, stored in a retrieval system, or transmitted in any form or by any means, electronic, mechanical, photocopying, microfilming, recording, or otherwise without written permission from the Publisher. Methods in Molecular Biology™ is a trademark of The Humana Press Inc.

All authored papers, comments, opinions, conclusions, or recommendations are those of the author(s), and do not necessarily reflect the views of the publisher.

This publication is printed on acid-free paper. ∞
ANSI Z39.48-1984 (American Standards Institute) Permanence of Paper for Printed Library Materials.

Cover design by Patricia F. Cleary.

Production Editor: Mark J. Breaugh.

For additional copies, pricing for bulk purchases, and/or information about other Humana titles, contact Humana at the above address or at any of the following numbers: Tel.: 973-256-1699; Fax: 973-256-8341; E-mail: humana@humanapr.com; Website: http://humanapress.com

Photocopy Authorization Policy:
Authorization to photocopy items for internal or personal use, or the internal or personal use of specific clients, is granted by Humana Press Inc., provided that the base fee of US $10.00 per copy, plus US $00.25 per page, is paid directly to the Copyright Clearance Center at 222 Rosewood Drive, Danvers, MA 01923. For those organizations that have been granted a photocopy license from the CCC, a separate system of payment has been arranged and is acceptable to Humana Press Inc. The fee code for users of the Transactional Reporting Service is [0-89603-618-9/02 $10.00 + $00.25].

Printed in the United States of America. 10 9 8 7 6 5 4 3 2 1

Library of Congress Cataloging in Publication Data

DNA methylation protocols / edited by Ken I. Mills and Bernie H. Ramsahoye.
  p. cm. -- (Methods in molecular biology ; v. 200)
  Includes bibliographical references and index.
  ISBN 0-89603-618-9 (alk. paper)
  1. DNA--Methylation--Laboratory manuals. I. Mills, Ken I. II. Ramsahoye, Bernie H.
III. Series.

QP624.5.M46 D63 2002
572.8'6--dc21

2001051654

Preface

There has been a marked proliferation in the number of techniques available for studying methylation, and the field promises to be remarkably vibrant over the next decade. *DNA Methylation Protocols* covers the new and exciting techniques currently available in the analysis of DNA methylation and methylases. The techniques presented in this book should provide the researcher with most of the tools necessary for studying methylation at the global level and at the level of the sequence. In particular, techniques useful for identifying genes that might be aberrantly methylated in cancer and aging are well-represented. The book is not intended to be an exhaustive account of all the techniques available, but does cover most of the recent substantive breakthroughs in methodology.

Ken I. Mills, PhD
Bernard H. Ramsahoye, MD, PhD

Contents

Preface v
Contributors ix

 1 Overview
   Ken I. Mills and Bernard H. Ramsahoye 1
 2 Nearest-Neighbor Analysis
   Bernard H. Ramsahoye 9
 3 Measurement of Genome-Wide DNA Cytosine-5 Methylation
   by Reversed-Phase High-Pressure Liquid Chromatography
   Bernard H. Ramsahoye 17
 4 Methylation Analysis by Chemical DNA Sequencing
   Piroska E. Szabó, Jeffrey R. Mann, and Gerd P. Pfeifer 29
 5 Methylation-Sensitive Restriction Fingerprinting
   Catherine S. Davies 43
 6 Restriction Landmark Genome Scanning
   Joseph F. Costello, Christoph Plass, and Webster K. Cavenee ... 53
 7 Combined Bisulfite Restriction Analysis (COBRA)
   Cindy A. Eads and Peter W. Laird 71
 8 Differential Methylation Hybridization Using CpG Island Arrays
   Pearlly S. Yan, Susan H. Wei, and Tim Hui-Ming Huang 87
 9 Methylated CpG Island Amplification for Methylation Analysis
   and Cloning Differentially Methylated Sequences
   Minoru Toyota and Jean-Pierre J. Issa 101
10 Isolation of CpG Islands Using a Methyl-CpG Binding Column
   Sally H. Cross 111
11 Purification of MeCP2-Containing Deacetylase
   from *Xenopus laevis*
   Peter L. Jones, Paul A. Wade, and Alan P. Wolffe 131
12 DNA-Methylation Analysis
   by the Bisulfite-Assisted Genomic Sequencing Method
   Petra Hajkova, Osman El-Maarri, Sabine Engemann,
   Joachim Oswald, Alexander Olek, and Jörn Walter 143
13 Measuring DNA Demethylase Activity In Vitro
   Moshe Szyf and Sanjoy K. Bhattacharya 155

14 Extracting DNA Demethylase Activity from Mammalian Cells
   Moshe Szyf and Sanjoy K. Bhattacharya 163

Index 177

Contributors

Sanjoy K. Bhattacharya • *Department of Pharmacology and Therapeutics, McGill University, Montreal, Canada*
Webster K. Cavenee • *Ludwig Institute for Cancer Research, University of California at San Diego, La Jolla, CA*
Joseph F. Costello • *Department of Neurological Surgery, UCSF Brain Tumor Research Center, San Francisco, CA*
Sally H. Cross • *MRC Human Genetics Unit, Western General Hospital, Edinburgh, UK*
Catherine S. Davies • *Department of Medical Biochemistry, University of Wales College of Medicine, Cardiff, UK*
Cindy A. Eads • *Department of Biochemistry and Molecular Biology, USC Norris Comprehensive Cancer Center, Los Angeles, CA*
Osman El-Maarri • *Max-Planck-Institute for Molecular Genetics, Berlin, Germany*
Sabine Engemann • *Max-Planck-Institute for Molecular Genetics, Berlin, Germany*
Petra Hajkova • *Max-Planck-Institute for Molecular Genetics, Berlin, Germany*
Tim Hui-Ming Huang • *Department of Pathology and Anatomical Sciences, Ellis Fischel Cancer Center, University of Missouri-Columbia, Columbia, MO*
Jean-Pierre J. Issa • *Graduate School of Biomedical Sciences, MD Anderson Cancer Center, Houston, TX*
Peter L. Jones • *Department of Molecular Embryology, National Institute of Child Health and Human Development, Bethesda, MD*
Peter W. Laird • *Department of Surgery and Biochemistry and Molecular Biology, USC Norris Comprehensive Cancer Center, Los Angeles, CA*
Jeffrey R. Mann • *Department of Biology, Beckman Research Institute of the City of Hope, Duarte, CA*
Ken I. Mills • *Department of Haematology, University of Wales College of Medicine, Cardiff, UK*
Alexander Olek • *Max-Planck-Institute for Molecular Genetics, Berlin, Germany*

JOACHIM OSWALD • *Max-Planck-Institute for Molecular Genetics, Berlin, Germany*

CHRISTOPH PLASS • *Division of Cancer Genetics, The Ohio State University, Columbus, OH*

GERD P. PFEIFER • *Department of Biology, Beckman Research Institute of the City of Hope, Duarte, CA*

BERNARD H. RAMSAHOYE • *Department of Haematology, Western General Hospital, Edinburgh, UK*

PIROSKA E. SZABÓ • *Department of Biology, Beckman Research Institute of the City of Hope, Duarte, CA*

MOSHE SZYF • *Department of Pharmacology and Therapeutics, McGill University, Montreal, Canada*

MINORU TOYOTA • *Graduate School of Biomedical Sciences, MD Anderson Cancer Center, Houston, TX*

PAUL A. WADE • *Department of Molecular Embryology, National Institute of Child Health and Human Development, Bethesda, MD*

JÖRN WALTER • *Max-Planck-Institute for Molecular Genetics, Berlin, Germany*

SUSAN H. WEI • *Department of Pathology and Anatomical Sciences, Ellis Fischel Cancer Center, University of Missouri-Columbia, Columbia, MO*

ALAN P. WOLFFE • *Department of Molecular Embryology, National Institute of Child Health and Human Development, Bethesda, MD*

PEARLLY S. YAN • *Department of Pathology and Anatomical Sciences, Ellis Fischel Cancer Center, University of Missouri-Columbia, Columbia, MO*

1

Overview

Ken I. Mills and Bernard H. Ramsahoye

In the last decade great strides have been made in understanding the molecular biology of the cell. The entire sequence of the human genome, and the entire genomes of a number of other organisms and microorganisms, are now available to researchers on the World Wide Web. As we enter the postgenome era, research efforts will increasingly focus on the mechanisms that control the expression of genes and the interactions between proteins encoded by the genomic DNA. Most of what we know about DNA methylation in mammals indicates that it is likely to be part of a system affecting chromatin structure and transcriptional control. As such, mammalian DNA methylation has traditionally attracted intense research interest from scientists in the fields of Development and Cancer biology. The recent discovery that two human diseases, ICF syndrome *(1)* (Immunodeficiency, Centromeric region instability, Facial abnormalities) and Rett syndrome *(2)*, a form of X-linked mental retardation, are caused by mutations in genes coding for a methyltransferase and a methyl-CpG binding protein, respectively, has broadened and intensified interest further. This book has been compiled in the hope that it will be a useful technical manual for those in the field of DNA methylation. What follows is a brief review of key facts and developments in the field in the hope that, for the uninitiated, this will help to set the technical chapters in context.

The DNA of most organisms is modified by the postsynthetic addition of a methyl group to carbon 5 of the cytosine ring. Although in prokaryotes other forms of methylation also exist (cytosine-N4, adenine-N6), DNA methylation in mammals is restricted to cytosine-5 and occurs almost exclusively within the dinucleotide sequence CpG. In mammalian DNA approx 80% of all CpG dinucleotides are methylated and the overall frequency of CpG is five times lower than expected given the frequencies of cytosine and guanine. The reason for this is thought to be the continued spontaneous hydrolytic deamination of 5-methylcytosine (at CpG) to thymine over the course of evolution. The regions of the genome that have been spared this deamination are those that are not ordinarily methylated. These regions are known as CpG islands and correspond with the promoter regions of more than half of all genes. Hydrolytic deamination of 5-methylcytosine to thymine continues to have an impact on cell biology and is of particular relevance in carcinogenesis. Indeed, 5-methylcytosine to thymine transitions are by far the most common form of mutation seen in cancer, accounting for at least 30% of the mutations described in the p53 gene *(3)*.

There has been a rapid expansion in the number of enzymes known to catalyze (or likely to catalyze) the cytosine-5 methylation reaction (the DNA cytosine-5 methyltransferase). The first mammalian methyltransferase to be described is now known as DNA methyltransferase 1 (Dnmt1) *(4)*. This enzyme is most probably responsible for maintaining the methylation states of sites through cell division. Dnmt1 is thought to be part of the replication machinery, being tethered to Proliferation Cell Nuclear Antigen (PCNA) through its N terminus *(5)*. It is the affinity of Dnmt1 for hemi-methylated DNA and the ability of Dnmt1 to restore full methylation to the hemi-methylated sites that arises as a result of semi-conservative replication, that ensures that methylation patterns are maintained once established. Dnmt2, a putative cytosine-5 methyltransferase based on sequence homology with other cytosine-5 methyltransferases, has not yet been shown to be active in vitro or in vivo *(6)*. The more recently discovered and related enzymes Dnmt3a and Dnmt3b, are highly expressed in embryonic cells and have *de novo* methyltransferase rather than maintenance methyltransferase activity *(7)*. That is to say, these enzymes are able to establish methylation on one or both strands at sites that were previously completely unmethylated. The Dnmt3a and Dnmt3b enzymes are major players in restoring methylation levels in the post-implantation embryo after global pre-implantation demethylation *(1)*.

As well as hastening the discovery of the methyltransferases, the sequencing effort has also accelerated the discovery of proteins that bind to methylated DNA. Since the first papers demonstrating methyl-CpG binding activity (MeCP) in nuclear extracts *(8,9)* it is now known that this activity (known as MeCP1) may result from the binding of different proteins in different cell types. The first of the methyl-CpG binding proteins to be characterized, methyl CpG binding protein 2 (MeCP2) *(10)* and four other proteins discovered by database homology searching using the methyl-CpG binding domain (MBD) of MeCP2 as bait (MBD1, MBD2, MBD3, and MBD4) *(11)*, are likely to confer at least some of the effects of DNA methylation. MBD2 is one of the methyl-CpG binding proteins responsible for MeCP1 activity *(12)*. In vitro experiments have demonstrated that MeCP2, MBD1, and MBD2 are likely to be involved in transcriptional repression through changes in the chromatin structure *(12–14)*. These proteins are components of co-repressor complexes containing histone deacetylase. They probably target the complexes to the methylated DNA by virtue of their methyl-CpG binding domains. MBD2 and MBD3 have also been shown to be core components of the NuRD chromatin remodeling complex *(15,16)*. MBD4, which turns out to be a thymidine N-glycosylase, is involved in the repair of G:T base-pair mismatches *(17)*. These mismatches may arise when 5-methylcytosine mutates to thymine by hydrolytic deamination.

The Role of DNA Methylation

Gene knock-out studies indicate that CpG methylation in mammals is an indispensable process. Targeted deletion of Dnmt1 results in a marked reduction in the level of DNA methylation in embryonic stem cells (ES) (25–30% of wild-type levels) and is lethal early in embryogenesis *(18)*. Combined deletion of Dnmt3a and Dnmt3b is also lethal at a very early stage, the resultant post-implantation embryos being highly demethylated *(1)*. These studies demonstrate the importance of DNA methylation in development but precisely why it is important is still not clear. The idea that methylation somehow orchestrates changes in chromatin structure during normal development is not well-supported by experimental evidence. Importantly, with the exception of the relative handful of imprinted genes and the genes on the inactive X chromosome in females, the CpG islands located in the promoter regions of genes are usually unmethylated irrespective of the expression status of the gene *(19)*. Hence promoter methylation does not appear to be a normal control mechanism in the expression of most CpG island-containing genes, even when they exhibit tissue-specific expression patterns. Genes containing non-CpG island promoters may exhibit a relationship between promoter methylation and transcriptional downregulation. However, some have argued that hypomethylation frequently follows transcriptional activation in the tissue-specific expression of genes. Therefore, the link between methylation and gene quiescence may not be causal.

In the case of imprinted genes and genes on the inactive X chromosome, methylation may have a more direct involvement in transcriptional control *(20,21)*. In both situations, only one of the parental alleles is active and the alleles are differentially methylated. Methylation of the inactive allele at the CpG island promoter probably has the effect of altering chromatin structure sufficiently to deny access to transcription factors that are otherwise available to the promoter of the active unmethylated allele. In the case of X chromosome inactivation in females, inactivation is initiated by *Xist* RNA *(22)*. This coats the X chromosome in *cis* and leads to its inactivation. However, CpG island methylation has emerged as an essential mechanism involved in the maintenance of the inactive state in the embryo. Interestingly, in the visceral endoderm which is derived from the extra-embryonic lineage and where X inactivation is not random but is imprinted (the paternal X is silent because the paternal Xist is unmethylated and active), DNA methylation appears not to be essential for maintaining X inactivation *(23)*.

DNA methylation is substantially deregulated in cancer. Global hypomethylation has been described by numerous authors in many tumors but the phenomenon is not universal *(24,25)*. Paradoxically, the hypermethylation of CpG island promoters is also well-described in cancer as well as in aging *(26)*. The propensity of CpG islands to become hypermethylated has led to the hypothesis that DNA methylation could provide an alternative (epigenetic) mechanism to loss of function of a tumor-suppressor gene. Hence, loss of function could be due to methylation of both alleles, mutation of both alleles, or a combination of methylation and mutation affecting individual alleles. While an increasing number of studies continue to highlight CpG island hypermethylation of tumor-suppressor genes in cancer, evidence that the relationship is causal falls short of proof. There is still the possibility that the genes found to be methylated in cancer and in cancerous cell lines are those that have been silenced by another mechanism, the methylation seen merely reinforcing a quiescent state.

Evidence from the two clinical syndromes recently linked to DNA methylation may offer further insights into the function of DNA methylation in mammals. The rare autosomal recessive syndrome ICF (Immunodeficiency, Centromeric region instability, Facial abnormalities) is now known to be due to DNMT3B deficiency *(1,27)*. A remarkable cytogenetic feature of this syndrome is the failure of the pericentromeric heterochromatic regions of chromosomes 1, 9, 16 to condense in metaphase chromosome preparations. This is similar to the situation observed when cells are treated with the demethylating agent 5-azacytidine. Consistent with this, the satellite II and III DNA is substantially demethylated in ICF syndrome. Loss of methylation of CpG islands on the inactive X chromosome is also seen, and this seems to result in derepression of transcription *(28)*. Mouse ES cells deficient in Dnm3b also have markedly demethylated minor satellite DNA. As DNMT3B deficiency leads to a widespread defect in DNA methylation throughout the genome, albeit with a certain predilection for satellite DNA sequences, it is intriguing that the phenotype induced should be that of Chromosomal instability, Facial abnormalities and Immune deficiency. Some chromosomal instability is not entirely unexpected because global demethylation induced by Dnmt1 deficiency in mouse ES cells has been shown to increase genome instability. However the mechanism leading to this instability is still unclear. Full characterization of the immunodeficiency in ICF patients may also help to reveal novel mechanisms involving DNA methylation.

In the case of the X-linked Rett syndrome, the establishment of MeCP2 mutations as the cause also prompts some reassessment of the cellular function of this methyl-CpG binding protein *(2)*. Prior to the detection of MeCP2 mutations in this syndrome, it would have been reasonable to assume that, as MeCP2 is widely expressed in somatic tissues and binds to methylated DNA, its function might have been crucial in many different tissues. Deficiency might have been expected to lead to global defects in transcriptional control and a phenotype apparent in many cell types and systems. However MeCP2 deficiency in Rett syndrome has a more specific phenotype than might have been predicted, leading to a neurodevelopmental defect and mental retardation in females. Could it be that there is some functional redundancy of the methyl-CpG binding proteins in all tissues except brain? Or does MeCP2 have some other function in promoting cognitive development, related or unrelated to its activity as a methyl-CpG binding protein?

DNA methylation promises to be a vibrant field over the next decade. There has been a marked proliferation in the number of techniques available for studying methylation. The techniques presented in this book should provide the researcher with most of the tools necessary for studying methylation at the global level and at the level of the sequence. In particular, techniques useful for identifying genes that might be aberrantly methylated in cancer and aging are well-represented. The book is not intended to be an exhaustive account of all the techniques available, but does cover most of the recent substantive breakthroughs in methodology. Established techniques, such as Southern hybridization of size-fractionated DNA digested with methylation-sensitive restriction enzymes, are not covered, but have been well-described elsewhere *(29)*.

References

1. Okano, M., Bell, D. W., Haber, D. A., and Li, E. (1999) DNA methyltransferases Dnmt3a and Dnmt3b are essential for de novo methylation and mammalian development. *Cell* **99**, 247–257.
2. Amir, R. E., Van den Veyver, I. B., Tran, C. Q., Francke, U., and Zoghbi, H. Y. (1999) Rett syndrome is caused by mutations in X-linked MECP2, encoding methyl-CpG-binding protein 2. *Nat. Genet.* **23**, 185–188.
3. Cooper, D. N. and Krawczak, M. (1990) The mutational spectrum of single base-pair substitutions causing human genetic disease: patterns and predictions. *Human Genet.* **85**, 55–74.
4. Bestor, T. H. and Ingram, V. M. (1983) Two DNA methyltransferases from murine erythroleukaemia cells: purification, sequence specificity, and mode of interaction with DNA. *Proc. Natl. Acad. Sci. USA* **80**, 5559–5563.

5. Chuang, L. S., Ian, H. I., Koh, T. W., Ng, H. H., Xu, G., and Li, B. F. (1997) Human DNA-(cytosine-5) methyltransferase-PCNA complex as a target for p21WAF1. *Science* **277,** 1996–2000.
6. Okano, M., Xie, S., and Li, E. (1998) Dnmt2 is not required for de novo and maintenance methylation of viral DNA is embryonic stem cells. *Nucleic Acids Res.* **26,** 2536–2540.
7. Okano, M., Xie, S., and Li, E. (1998) Cloning and characterisation of a family of novel mammalian DNA (cytosine-5) methyltransferases. *Nature Genet.* **19,** 219–220.
8. Meehan, R. R., Lewis, J. D., McKay, S., Kleiner, E. L., and Bird, A. P. (1989) Identification of a mammalian protein that binds specifically to DNA containing methylated CpGs. *Cell* **58,** 499–507.
9. Boyes, J. and Bird, A. (1991) DNA methylation inhibits transcription indirectly via a methyl-CpG binding protein. *Cell* **64,** 1123–1134.
10. Lewis, J. D., Meehan, R. R., Henzel, W. J., Maurer-Fogy, I., Klein, F., and Bird, A. (1996) Purification, sequence and cellular localisation of a novel chromosomal protein that binds to methylated DNA. *Cell* **69,** 905–914.
11. Hendrich, B. and Bird, A. (1998) Identification and characterisation of a family of mammalian methyl-cpG binding proteins. *Mol. Cell. Biol.* **18,** 6538–6547.
12. Ng, H. H., Zhang, Y., Hendrich, B., Johnson, C. A., Turner, B. M., Erdjument-Bromage, H., et al. (1999) MBD2 is a transcriptional repressor belonging to the MeCP1 histone deacetylase complex. *Nat. Genet.* **23,** 58–61.
13. Ng, H. H., Jeppesen, P., and Bird, A. (2000) Active repression of methylated genes by the chromosomal protein MBD1. *Mol. Cell. Biol.* **20,** 1394–1406.
14. Nan, X., Ng, H.-H., Johnson, C. A., Laherty, C. D., Turner, B. M., Eisenman, R. N., and Bird, A. (1998) Transcriptional repression by the methyl-CpG-binding protein MeCP2 involves a histone deacetylase complex. *Nature* **393,** 386–389.
15. Zhang, Y., Ng, H. H., Erdjument-Bromage, H., Tempst, P., Bird, A., and Reinberg, D. (1999) Analysis of the NuRD subunits reveals a histone deacetylase core complex and a connection with DNA methylation. *Genes Dev.* **13,** 1924–1935.
16. Feng, Q. and Zhang, Y. (2001) The MeCP1 complex represses transcription through preferential binding, remodeling, and deacetylating methylated nucleosomes. *Genes Dev.* **15,** 827–832.
17. Hendrich, B., Hardeland, U., Ng, H. H., Jiricny, J., and Bird, A. (1999) The thymine glycosylase MBD4 can bind to the product of deamination at methylated CpG sites. *Nature* **401,** 301–304.
18. Li, E., Bestor, T. H., and Jaenisch, R. (1992) Targeted mutation of the DNA methyltransferase gene results in embryonic lethality. *Cell* **69,** 915–926.
19. Bird, A. (1992) The essentials of DNA methylation. *Cell* **70,** 5–8.
20. Li, E., Beard, C., and Jeanisch, R. (1993) Role of DNA methylation in genomic imprinting. *Nature* **366,** 362–365.
21. Beard, C., Li, E., and Jaenisch, R. (1995) Loss of methylation activates Xist in somatic but not in embryonic cells. *Genes Dev.* **9,** 2325–2334.

22. Clemson, C. M., McNeil, J. A., Willard, H. F., and Lawrence, J. B. (1996) XIST RNA paints the inactive X chromosome at interphase: evidence for a novel RNA involved in nuclear/chromosome structure. *J. Cell Biol.* **132,** 259–275.
23. Sado, T., Fenner, M. H., Tan, S. S., Tam, P., Shioda, T., and Li, E. (2000) X inactivation in the mouse embryo deficient for Dnmt1: distinct effect of hypomethylation on imprinted and random X inactivation. *Dev. Biol.* **225,** 294–303.
24. Antequera, F., Boyes, J., and Bird, A. (1990) High levels of de novo methylation and altered chromatin structure at CpG islands in cell lines. *Cell* **62,** 503–514.
25. Jones, P. A. and Laird, P. W. (1999) Cancer epigenetics comes of age. *Nat. Genet.* **21,** 163–167.
26. Issa, J. P., Ottaviano, Y. L., Celano, P., Mamilton, S. R., Davidson, N. E., and Baylin, S. B. (1994) Methyation of the oestrogen receptor CpG island links ageing and neoplasia in human cancer. *Nature Genet.* **7,** 536–540.
27. Xu, G. L., Bestor, T. H., Bourc'his, D., Hsieh, C. L., Tommerup, N., Bugge, M., et al. (1999) Chromosome instability and immunodeficiency syndrome caused by mutations in a DNA methyltransferase gene. *Nature* **402,** 187–191.
28. Hansen, R. S., Stoger, R., Wijmenga, C., Stanek, A. M., Canfield, T. K., Luo, P., et al. (2000) Escape from gene silencing in ICF syndrome: evidence for advanced replication time as a major determinant. *Human Mol. Genet.* **9,** 2575–2587.
29. Sambrook, J., Fritsch, E. F., and Maniatis, T. (1989) *Molecular Cloning: A Laboratory Manual.* Cold Spring Harbor Laboratory Press, Cold Spring Harbor, New York.

2

Nearest-Neighbor Analysis

Bernard H. Ramsahoye

1. Introduction

Nearest-neighbor analysis can be used to identify the 3′ nearest neighbors of 5mC residues in DNA *(1,2)*. It can also be used to measure the level of methylation of a specific methylated dinucleotide in DNA. Typically, in the case of mammalian DNA, this means quantifying the degree of methylation at CpG dinucleotides. It has the added advantage of being applicable to small samples of the order of 1 microgram of genomic DNA. The only drawback is that it is a radioactive technique and the appropriate facilities and techniques for handling radioactive substances must be available.

1.1. Outline of the Procedure

DNA is digested with a restriction enzyme and labeled at a restriction enzyme cut site with Klenow fragment of DNA polymerase I and a $[\alpha\text{-}^{32}P]$ dNTP. After digestion of the labeled DNA to deoxyribonucleotide 3′-monophosphates using a combination of an exonuclease and an endonuclease, the radiolabeled 5′ α-phosphate of the $[\alpha\text{-}^{32}P]$ dNTP will appear as the 3′-phosphate of the nucleotide (X) that was immediately 5′ it in the DNA (its nearest neighbor). As labeling is template dependent, the amount of the labeled nucleotide 3′-monophosphate in the digest reflects the frequency of a dinucleotide (XpN) in the DNA. The technique of labeling cut sites by fill-in reaction as described here is a modification of the nick-labeling nearest-neighbor analysis technique first published by Gruenbaum et al. *(3)*. In the author's experience, the original technique of using DNase I to nick the DNA and the DNA polymerase I holoenzyme to label the DNA by nick translation gives less reproducible results than the fill-in method using Klenow fragment of DNA polymerase I.

From: *Methods in Molecular Biology, vol. 200: DNA Methylation Protocols*
Edited by: K. I. Mills and B. H. Ramsahoye © Humana Press Inc., Totowa, NJ

1.2. Quantification of CpG Methylation in Mammalian DNA

If $Mbo(\backslash GATC)$ is used to cut the DNA and $[\alpha\text{-}^{32}P]$ dGTP is used to label it, after digestion of the labeled DNA to deoxyribonucleotide 3′-monophosphates, the quantities of labeled 5mdCp, dCp, Tp, dGp, and dAp reflect the relative frequencies of the dinucleotides 5mdCp**G**, dCp**G**, Tp**G**, dGp**G**, and dAp**G** at *MboI* cut sites in the DNA.

1.3. Quantification of Non-CpG and CpG Methylation

If the sequence context of cytosine-5 methylation is unknown it may not be wise to assume that it is at CpG. When methylation is present in sequences other than CpG the DNA can be cut with *FokI* (GGATGN$_{9-13}$) and labeled separately with each of the 4 $[\alpha\text{-}^{32}P]$ dNTPs. All dinucleotides containing a 5′ 5-methylcytosine should be detectable using this approach. The rational for using *FokI* here is that even if the methylation only occurred within a specific sequence in the DNA (a 4–6 base recognition sequence) there would be an approx 1 in 1000 chance that such sites would also have a *FokI* recognition sequence upstream of the methylated site. Thus if methylation occurred consistently within a 4–6 base sequence context it should be detectable using this technique, albeit at low level. It should be noted that there is a hypothetical possibility that methylation could be missed if the pattern of methylation in the sample was such that it always arose 9–13 bases downstream of a specific sequence in the DNA. Using *FokI* in this instance might positively exclude the detection of these methylated sites. Also, if the genome of the organism was particularly small (of the order of 10^6 bases) and methylation occurred within a specific 5 or 6 base sequence only, too few methylated sites might be present downstream of a *FokI* site to reliably allow their detection using this enzyme.

2. Materials
2.1. Reagents

1. High molecular-weight DNA.
2. A restriction enzyme that reliable cuts cytosine-5 methylated DNA leaving a 5 overhang, e.g., *FokI* for the detection of 5mC at 5mCpN, *MboI* for the detection of 5mC at 5mCpG and *MvaI* for detecting methylation of the internal cytosine in the sequence CC\WGG.
3. $[\alpha\text{-}^{32}P]$ dNTP (3000Ci/mmol, Amersham Pharmacia Biotech).
4. Klenow fragment of DNA polymerase I + labeling buffer (Amersham Pharmacia Biotech).
5. Micrococcal nuclease (P6752, Sigma).
6. Calf spleen phosphodiesterase (Worthington Biochemical Corporation).

Nearest-Neighbor Analysis

7. Micrococcal nuclease/spleen phosphodiesterase digestion buffer: 15 mM CaCl$_2$, 100 mM Tris-HCl.
8. 0.2 M ethylenediaminetetraacetic acid (EDTA) (Sigma).
9. Sephadex G50 spin columns (available from Roche).
10. Solution A: 66 volumes isobutyric acid: 18 vol water: 3 vol 30% ammonia solution.
11. Solution B: 80 volumes saturated ammonium sulphate: 18 vol 1 M acetic acid: 2 vol isopropanol.

2.2. Equipment

1. Radioactivity laboratory equipped with protective screens.
2. Disposable gloves should be worn at all times.
3. Water bath set at 15°C.
4. Hot block set at 37°C.
5. DNA vacuum drier (e.g., Speed Vac).
6. Thin-layer chromatography (TLC) developing tanks.
7. Glass-backed 20 cm × 20 cm cellulose TLC plates.
8. X-ray film.
9. X-ray cassettes.
10. Developer.
11. Phosphorimager or scintillation counter.

3. Method
3.1. Estimation of Percent Methylation at CpG

1. Extract DNA from the tissue to be analyzed. Any of the standard methods can be used but the DNA should be high molecular weight and free of RNA. RNA should be removed by enzymatic hydrolysis with RNaseA and RnaseT1 (together) followed by recovery of the DNA by ethanol precipitation.
2. Digest 1 μg DNA with 10 units of *MboI* at 37°C overnight.
3. Heat-inactivate the enzyme (70°C for 20 min).
4. Precipitate the DNA in ethanol, pellet by centrifugation, and re-dissolve the DNA in 10 μL of water. Whilst the DNA is re-dissolving, prepare an appropriate number of Sephadex G50 columns in order that these are ready for use on completion of the labeling step.
5. Add 3 μL [α-32p]dGTP (30 μCi), 1.5 μL 10X labeling buffer, and 0.5 μL Klenow on ice.
6. Incubate for 15 min at 15°C.
7. Add 2 μL 0.2 M EDTA to terminate the reaction.
8. Carefully transfer the labeling mixture to the top of a Sephadex G50 spin column.
9. Centrifuge at 1100g for 4 min collecting the flow through in a 1.5 mL polypropylene tube.

10. Dry down the labeled DNA in a DNA speed vac.
11. Digest the DNA in a volume of 7 µL (5 µL micrococcal nuclease digestion buffer, 1 µL [0.2 units] micrococcal nuclease and 1 µL [2 µg] spleen phosphodiesterase. The digest should be complete after 4 h at 37°C.
12. Proceed to TLC or freeze the sample at –20°C until ready to proceed to TLC.

3.2. Preparation of Sephadex G50 Columns

1. Sephadex G50 columns can be purchased from commercial suppliers (Roche). They can also be prepared more cheaply in house using 1-mL syringes, swollen Sephadex G50, and glass wool (to plug the syringe and prevent escape of sephadex during centrifugation).
2. To prepare your own columns, roll a small amount of glass wool between a gloved finger and thumb and insert it into a 1-mL syringe using the syringe plunger. The amount of glass wool should be such that it is just sufficient to cover the exit hole of the syringe and prevent the escape of sephadex during centrifugation.
3. Pipet sephadex G50 slurry into the barrel of the syringe and fill to the brim. Insert the syringe into a 15-mL tube (Falcon).
4. Centrifuge at 1100g for 2 min to compact the G50 and expel the buffer.
5. Pipet more G50 slurry into the barrel (filling to the brim) and centrifuge again.
6. The compacted sephadex G50 is now ready for sample loading. The sample should be applied to the center of the column and a 1.5-mL polypropylene tube should be placed in the 15-mL Falcon tube to collect the elute after centrifugation.
7. Centrifuge the sample at 1100g for 4 min to separate the labeled DNA (which appears in the elute) from the free nucleotides (which are retained in the column).

3.3. Thin-Layer Chromatography

1. In the author's experience, TLC developing tanks designed to take more than two plates in near vertical positions give suboptimal separations in this application. Standard tanks that allow for allow a maximum of two plates to be developed at once, give improved separations as the plates can be positioned at a more favourable angle (**Fig. 1A**).
2. The DNA should be labeled to a high specific activity. Ordinarily the tube containing the digested ^{32}P-labeled DNA should read more than 2000 counts/s when placed up against a Geiger counter.
3. Using a 2-µL pipet, spot 0.3 µL of the digest onto a 20 × 20 cm glass-backed cellulose TLC plate 1.5 cm from the bottom right corner. Take care not to mark the cellulose in the process. The plate should be labeled with a pencil in the top left corner (**Fig. 1B**). The position (for application) can be marked lightly beforehand with a pencil. If the DNA is insufficiently labeled (there was too little DNA) then the sample may have to be applied repeatedly (with intervening

A

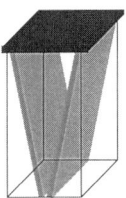

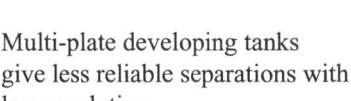

Multi-plate developing tanks give less reliable separations with less resolution

Standard developing tanks taking a maximum of 2 plates give reliable separations

B

Cellulose covered glass backed TLC plate

The sample is applied 1.5 cm from the bottom right corner of the TLC plate

Fig. 1. **(A)** TLC developing tanks. **(B)** Applying the sample to the TLC plate.

drying) to the same spot. This should be avoided if possible as it will inevitable detract from the resolution of the subsequent chromatography. Ideally a single 0.3 µL application should give a measurement of 500–1000 counts/s when a Geiger counter is held directly over it.
4. Make up solution A fresh prior to each use. The solution should be made up in a fume hood (isobutyric acid fumes are foul-smelling and toxic) and all subsequent chromatography should be carried out in the fume hood.
5. Pour 44 mL of solution A into a TLC developing tank complete with glass lid. Ensure that there is a good seal.
6. When the applied sample is dry, carefully place the TLC plate at an angle in the developing tank and replace the lid (**Fig. 1A**). Allow the plate to develop fully. This should take 12–18 h, the time taken being dependent on the ambient temperature. Separations are quicker but noticeably poorer in the summer months. If an elevated ambient temperature is a problem attempts should be made to carry out the chromatography in an air-conditioned room.
7. Once the plate is fully developed, remove it carefully and place it on absorbent paper (cellulose side uppermost) behind a radiation screen with the fume-hood extractor turned on. The plate will take about 4 h to dry thoroughly. Incomplete drying of the plate adversely affects the subsequent chromatography.

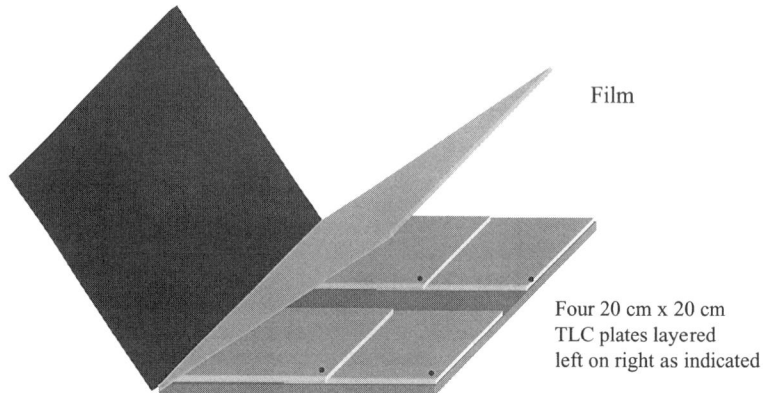

Fig. 2. Arrangement for stacking four TLC plates in a single autoradiography cassette.

8. The solution A in the developing tank should then be poured off into a container for solvent waste and the tank should then be washed out thoroughly in water (taking care not the splash the drying TLC plate).
9. It is preferable for all of these steps to be carried out in the fume hood (if equipped with a sink) as the residual isobutyric acid will leave a foul smell even in a well-ventilated room.
10. Once the TLC plate is dry, turn the plate through 90 degrees and subject the sample to the second dimension of chromatography using solution B.
11. When the second dimension is complete, remove the plate and dry again in the fume hood with the extractor on. Drying with the extractor on is essential as otherwise coarse crystallization of the ammonium sulphate leads to deterioration and flaking of the cellulose layer.
12. When drying is complete the plates can be analyzed by autoradiography or phosporimaging. In the case of autoradiography the labeled nucleotides can subsequently be quantified by scintillation. It is possible to fit four TLC plates in one 35×43 cm autoradiography cassette if they are stacked as indicated in **Fig. 2**. This saves on X-ray film and so is more economical. A 24-h exposure is usually sufficient to locate even low levels of methylation.
13. After developing the film, the autoradiograph is used to locate the position of the labeled nucleotides on the TLC plates (**Fig. 3**). Tracing paper is used to record the positions with a pencil, drawing a circle around each nucleotide. The tracing paper can then be applied directly to the plate and a pencil used to delineate the positions of the respective nucleotides on the plate.

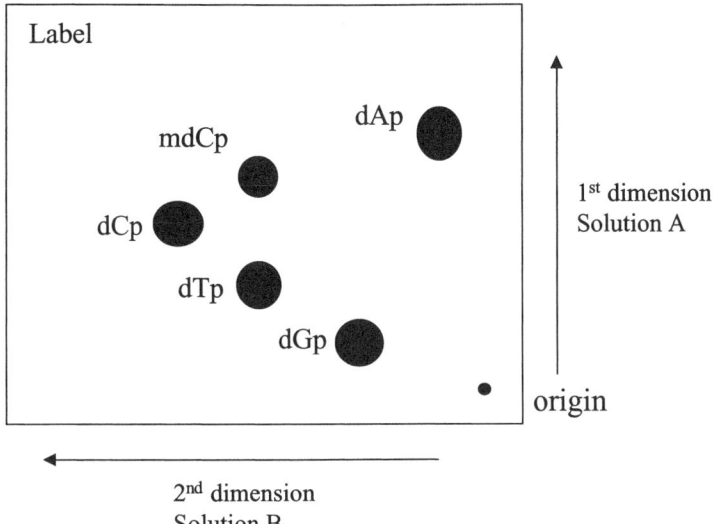

Fig. 3. Expected positions of nucleotide 3′-monophosphates after two dimensions of chromatography.

14. The cellulose can then be scraped off into scintillation vials using clean scalpel blades (being careful to recover all of the label and not to cross-contaminate other nucleotides).
15. The radioactivity in the cellulose can then be measured by scintillation after the addition of scintillant. Accurate and reproducible quantifications are only possible if great care is taken to ensure that all the cellulose is recovered without spillage.
16. When counting the relatively high-energy β emissions from ^{32}P, it is usually sufficient to record the scintillation data in units of counts per minute (CPM). This is because ^{32}P emissions are not significantly quenched in the circumstances described above. Construction of a quench curve and conversion of the data to disintegrations per minute (DPM) should not be necessary. The same would not apply if using the weaker emitting ^{33}P isotope in labeling reactions.

References

1. Lyko, F., Ramsahoye, B. H., and Jaenisch, R. (2000) DNA methylation in Drosophila melanogaster. *Nature* **408**, 538–540.
2. Ramsahoye, B. H., Biniszkiewicz, D., Lyko, F., Clark, V., Bird, A. P., and Jaenisch, R. (2000) Non-CpG methylation is prevalent in embryonic stem cells and may be mediated by DNA methyltransferase 3a. *Proc. Natl. Acad. Sci. USA* **97**, 5237–5242.
3. Gruenbaum, Y., Stein, R., Cedar, H., and Razin, A. (1981) Methylation of CpG sequences in eukaryotic DNA. *FEBS Lett.* **124**, 67–71.

3

Measurement of Genome-Wide DNA Cytosine-5 Methylation by Reversed-Phase High-Pressure Liquid Chromatography

Bernard H. Ramsahoye

1. Introduction

In health, approx 4% of cytosines are methylated. Tissues can vary in their levels of DNA methylation and the overall level is often reduced in malignancy *(1)*. The level of DNA methylation is usually obtained by chromatographic separation of the constituent nucleotide bases or their related deoxyribonucleotides or deoxyribonucleosides, and is usually represented as a fraction of total cytosine. Quantification of 5mC by chromatographic separation of deoxyribonucleotides has the advantage that, as deoxyribonucleotides can be easily distinguished from ribonucleotides, contamination of the DNA by RNA is less likely to cause error. This can be the case if 5mC is assayed after chemical hydrolysis of the DNA to bases.

The method outlined is a modification of the method of Kuo et al. *(2)* which was developed for the measurement of deoxyribonucleosides and later improved *(3)*. The chromatographic technique enables the complete separation of all five deoxyribonucleotides, making dephosphorylation of the nucleotides to nucleosides unnecessary. 5-methyl-2'-deoxycytidine-5'-monophosphate (5mdCMP) is measured as a proportion of total 2'-deoxycytidine-5'-monophosphates (5mdCMP + dCMP), and the technique is suitable for measuring the 5mdCMP content in 1 µg or more of DNA. Methods based on the measurement of nucleotide bases have been described elsewhere *(4,5)*.

From: *Methods in Molecular Biology, vol. 200: DNA Methylation Protocols*
Edited by: K. I. Mills and B. H. Ramsahoye © Humana Press Inc., Totowa, NJ

2. Materials
2.1. Enzymes
1. Ribonuclease A (RNase A).
2. Ribonuclease T1 (RNase T1).
3. Deoxyribonuclease I (DNase I).
4. Nuclease P1.

2.2. Buffers
1. Tris-EDTA (TE) 10 mM Tris-HCl, pH 8.0, 1 mM ethylenediaminetetraacetic acid (EDTA).
2. DNase I digestion buffer: 10 mM Tris-HCl, pH 7.2, 0.1 mM EDTA, 4 mM magnesium chloride.
3. 30 mM sodium acetate, pH 5.2.

2.3. Other Reagents
1. 10 mM Zinc sulphate.
2. 3 M Sodium acetate, pH 5.2.
3. Ethanol.
4. Phenol/chloroform/isoamyl alcohol, pH 8.0.

2.4. Solutions for High-Pressure Liquid Chromatography (HPLC)
1. 40% methanol.
2. 10% methanol.
3. 50 mM ammonium orthophosphate: this is made by dissolving 50 millimoles of diammonium orthophosphate in 1 L of 50 mM orthophosphoric acid with subsequent adjustment of the pH to 4.1 with 1 M orthophosphoric acid.

2.5. Nucleotide Standards
1. 2′-Deoxycytidine 5′-monophosphate (dCMP).
2. 5-Methyl-2′-deoxycytidine, 5′-Monophosphate (5mdCMP).
3. 2′-Deoxyguanosine 5′-monophosphate (dGMP).
4. 2′-Deoxyadenosine 5′-monophosphate (dAMP).
5. Thymidine 5′-monophosphate (TMP).

All enzymes, compounds, and solutions are available from Sigma.

3. Method
3.1. DNA Preparation

For the most accurate measurement of the constituent nucleotides, it is important to ensure that all RNA is removed from the DNA preparation.

Although ribonucleotides usually elute from the column with different retention times, their presence can sometimes interfere with measurement of the deoxyribonucleotides. DNA that has been prepared by conventional techniques such as phenol/chloroform extraction contains substantial amounts of RNA. Removal of RNA can be achieved by enzymatic hydrolysis with a combination of RNase A and RNase T1 followed by ethanol precipitation. The inclusion of RNase T1 is essential for total removal of the RNA (*see* **Note 1**).

1. Dissolve approx 50 µg DNA in 300 µL of 1X TE in a 1.5 mL polypropylene microcentrifuge tube. Add RNase A to a final concentration of 100 µg/mL and RNase T1 to a final concentration of 2,000 units/mL. Mix gently and incubate the solution at 37°C for 2 h.
2. Following the incubation, add an equal volume of phenol/chloroform/isoamyl alcohol and invert the tube several times to encourage mixing. Centrifuge for 2 min at 15,000 rpm in a bench-top microcentrifuge and gently remove the top aqueous layer containing the DNA into a clean tube by pipetting. Care should be taken not to carry over any phenol as this can interfere with the subsequent enzymatic hydrolysis as well as the chromatography. Contaminating phenol can be removed by extraction with ether.
3. Precipitate the DNA by adding 0.1 vol of 3 *M* sodium acetate and 2.5 vol of absolute ethanol. Recover the DNA by centrifugation and removal of the ethanol supernatant containing the hydrolyzed RNA. Wash the DNA pellet with 70% ethanol and resuspend in 100 µL deoxyribonuclease I (DNase I) digestion buffer.

3.2. DNA Hydrolysis

1. Follow **steps 1–3**, **Subheading 3.1.**
2. Add DNase I to a final concentration of 50 µg/mL and incubate at 37°C for 14 h.
3. Add 2 vol of 30 m*M* sodium acetate, pH 5.2, and zinc sulphate to a final concentration of 1 m*M*. Add Nuclease P1 to a final concentration of 50 µg/mL and incubate for a further 7 h at 37°C.
4. Removal of any solid debris from the DNA hydrolysate ensures that subsequent HPLC runs smoothly. To achieve this, the DNA hydrolysate can be filtered by centrifugation using a spin column with a 0.45 µm filter (Millipore).

To maximize the reproducibility of subsequent HPLC, samples should be injected in the same injection volume. The concentration of deoxyribonucleotides can be assessed by UV spectrophotometry at 260 nm and the concentration of the nucleotides adjusted with DNA digestion buffer (a solution containing 1 part DNase I digestion buffer to 2 parts 30 m*M* sodium acetate) so as to contain approx 5 µg of nucleotide in a 50 µL injection volume. The sample can be analyzed immediately or stored at –70°C until analysis.

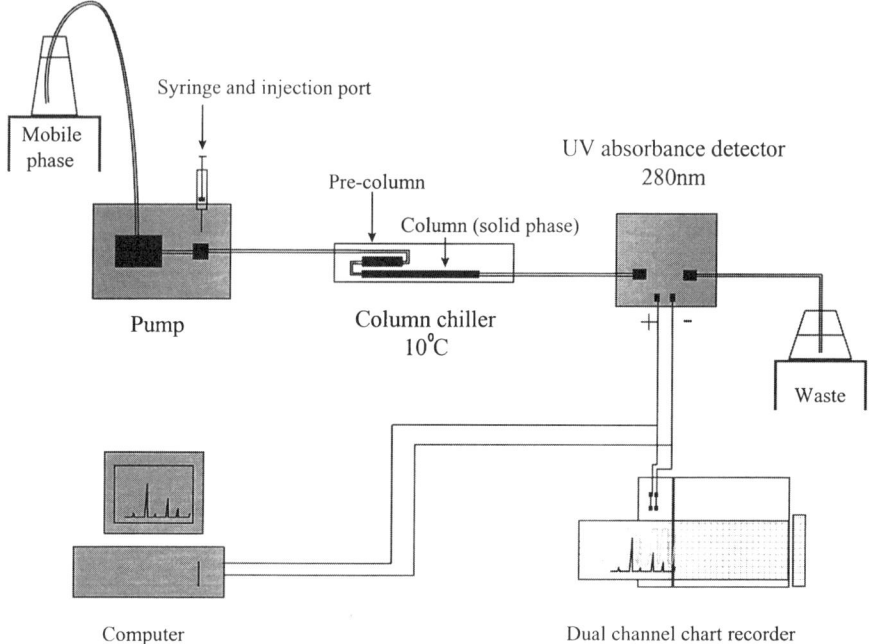

Fig. 1. HPLC apparatus.

3.3. Isocratic Reverse-Phase High-Pressure Liquid Chromatography (RP-HPLC)

The principle of RP-HPLC is that the components of a solution are separated by injecting them onto a column containing a nonpolar hydrocarbon chemically bonded onto the surface of rigid silica particles (solid phase). The components are then eluted from the column according to their solubility in a polar solution (mobile phase). Individual compounds are released from the column when specific volume of mobile phase has passed through. If the elution times of the compounds are sufficiently different, the compounds can be detected individually and quantified. The choice of mobile phase and solid phase depend on the application and make a marked difference to the efficiency of separation. In addition, variations in pH of the mobile phase and variations in ambient temperature also markedly affect separation. The mobile phase is delivered at high pressure and at a constant rate to the solid phase by means of a pump. The detection system is located down stream of the solid phase. An illustration of the HPLC apparatus is shown in **Fig. 1**.

For the separation of 2′-deoxyribonucleotide-5′-monophosphates (dNMPs), the following components and conditions are required:

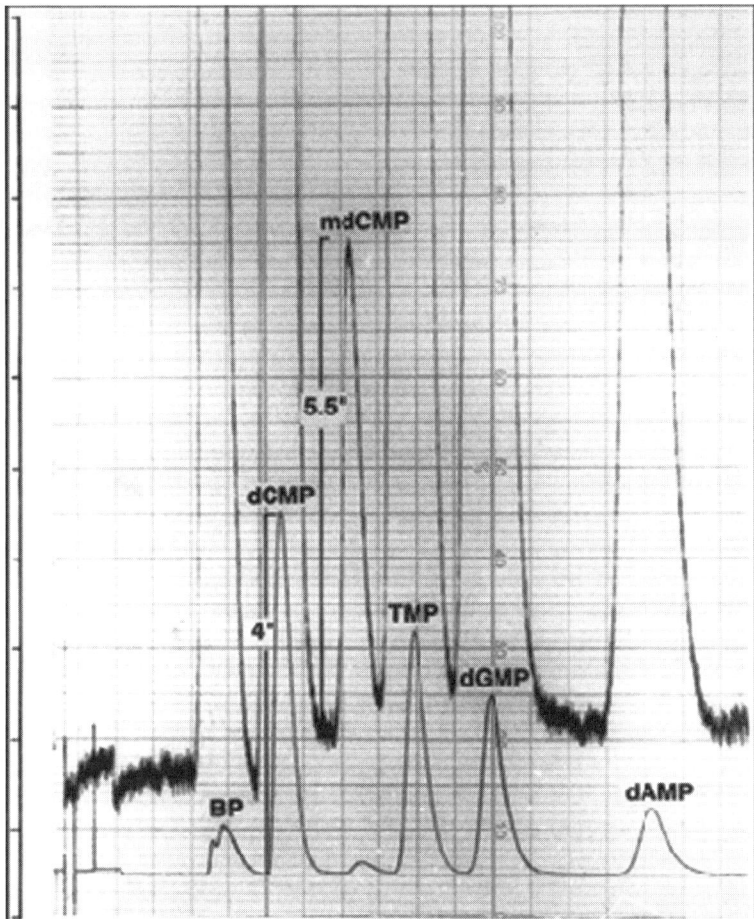

Fig. 2. Analysis of the methylation level in DNA extracted from human bone marrow mononuclear cells. A dual-channel chart recorder has been used to display the results. The separation was performed with the column at ambient temperature.

1. Mobile phase: 50 mM ammonium orthophosphate, pH 4.1. The mobile phase should be run at 1 mL/min, and should be filtered and degassed thoroughly before use.
2. Solid phase: 25 × 0.4 cm, 5 µm APEX ODS column (Jones Chromatography Limited, New Road, Hengoed, Mid Glamorgan, Wales, UK). Use of a pre-column is optional but will preserve the life of the column.
3. Column chiller: To improve reproducibility and aid separation, the column (and precolumn if present) should be chilled to 10°C. Chillers specifically designed for this purpose are available from Jones Chromatography. The benefit of chilling the column can be seen by comparing the peak separation in **Fig. 2** (column at ambient temperature) with that seen in **Fig. 3** (column chilled at 10°C).

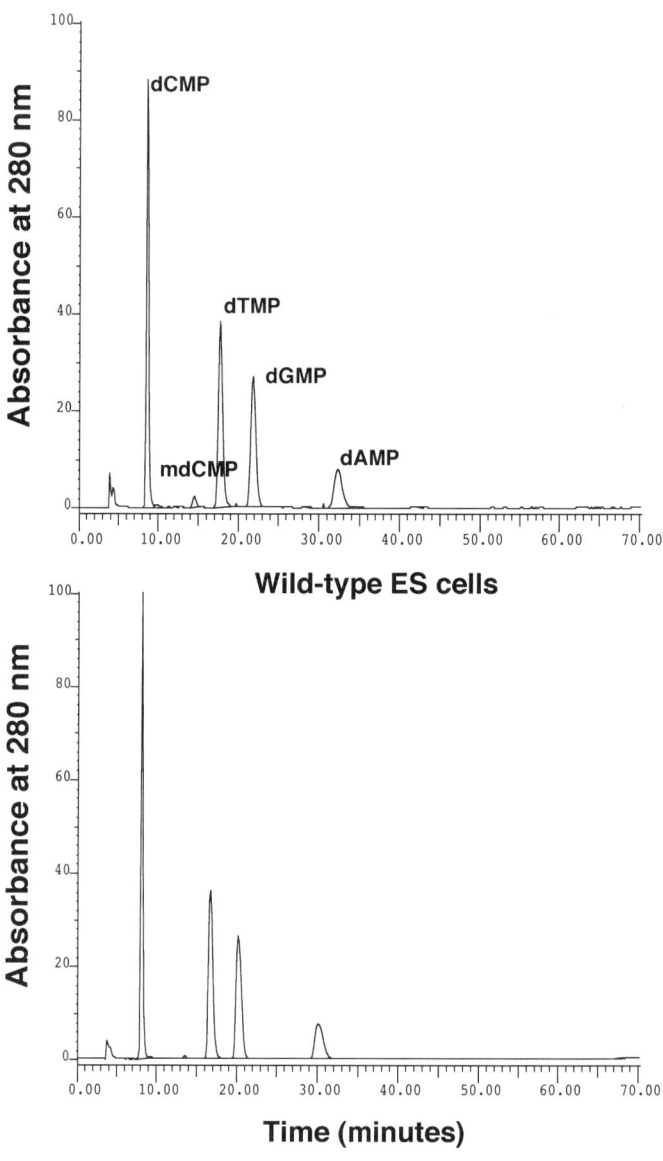

Fig. 3. Analysis of the DNA methylation level in wild-type mouse embryonic stem cells and DNA methyltransferase-deficient embryonic stem cells *(7)*. 5mdCMP in the mutant mice is much reduced but not absent. The column was chilled to 10°C, improving the separation between dNMP peaks (*see* **Fig. 2** for comparison) and the chromatogram was generated by Gilson 712 data acquisition and analysis software (*see* **Note 1**).

It is essential to ensure that the pump is functioning satisfactorily, at steady flow rates and constant pressure. Variations in flow rate will lead to unacceptable variations in the quantities of nucleotide detected, slowing of the rate past the detector leading to an increase in the area under the peak. Most systems include pressure gauges to indicate when the system is malfunctioning (*see* **Note 2**).

3.3.1. Nucleotide Detection

Nucleotides are best detected by a UV absorbance detector. Most modern detectors enable detection at any wavelength. Many detectors have two channels, allowing a single sample to be analyzed by two wavelengths of UV simultaneously and this may be useful for the identification of unknown compounds. In the present method, all nucleotides are analyzed at 280 nm, which is close to the λ_{max} of dCMP and 5mdCMP (272.7 and 278 nm, respectively).

Detectors are also able to detect at varying sensitivities (0.001–2 AUFS, absorbance unit full scale). The sensitivity of detection can be set in accordance with the quantity of DNA being analyzed. However, depending on the quality of the detector, at high sensitivity (low AUFS), the amount of background noise may be unacceptable leading to poor signal-to-noise ratios and poor coefficients of variation for peaks and peak ratios.

The UV absorbance detector measures the change in UV absorbance as the nucleotide passes the detector and converts this to an electrical signal that can be detected using a chart recorder or a computer system with the appropriate interface and software. Computer analysis of the absorbance-detector output is preferable as it is less subjective and more accurate. It also facilitates storage and manipulation of the data.

3.3.2. Computer Analysis

Using an appropriate interface, the UV absorbance data can be stored and analyzed using a variety of programs, the system used by the author being the Windows-based Gilson 712 software package (version 1.2).

3.3.3. Using a Chart Recorder

The main difficulty to overcome when using a chart recorder to measure small differences in % cytosine methylation is that when analyzing human DNA, the sizes of the 5mdCMP and dCMP peaks are markedly different, 5mC being approx 4% of total cytosine. Assessing a change in the size of the smaller peak compared with the larger peak is therefore subject to a greater error. This can be overcome by using a dual-channel chart recorder that has the facility to alter the gain (the voltage signal required to produce a full-scale deflection) for each channel separately. The output of the absorbance detector in connected to both channels of the chart recorder in parallel. The sensitivity of detection of the first channel can then be set so as to increase the size of the 5mdCMP peak and make it comparable to the height of the dCMP peak in the second channel. Peaks can then be measured either by peak height (trough to peak) or by measuring the peak area. The former is considerable less time-consuming than the latter. The relationship between nucleotide quantity and peak height is linear over a wide range, and measurement of peak heights is therefore both convenient and accurate for the assessment of nucleotide quantity. **Figure 2** shows an example of a typical chromatogram produced by this technique.

3.3.4. Using the HPLC

It is essential that the HPLC apparatus is not left in ammonium orthophosphate buffer when not in use as precipitation of the buffer can cause considerable damage to the system. After use, the system should be left in a solution of 40% methanol.

Prior to use and before equilibrating the HPLC column with the ammonium orthophosphate buffer, the column should be washed by sequential treatments with a solution of 40% methanol in water (1 mL/min for 30 min) followed by a solution of 10% methanol (1 mL/min for 30 min). The buffer can then be changed to 50 mM ammonium orthophosphate pH 4.1 (1 mL/min) and the system should be allowed to equilibrate for a further hour in this buffer before samples are loaded. At this point, the baseline signal from the absorbance detector should be level. When HPLC is completed, the system should be washed through with water followed by solutions of 10% and 40% methanol each for 30 min.

3.4. Quantification of dCMP and 5mdCMP Using a Computerized Data Acquisition

The simplest way of measuring the molar equivalents of dCMP and 5mdCMP is by measuring the peak area at 280 nm and dividing this value by the extinction coefficient for each nucleotide at 280nm. The extinction coefficients at pH 4.3 and 280 nm for dCMP and 5mdCMP have been determined by Sinsheimer to be 11.5×10^3 and 10.1×10^3 respectively *(6)*. 5mdCMP can then be expressed as a percentage of total dCMP according to the following formula:

$$\% cytosine\ methylation = 5mdCMP\ \frac{5mdCMP}{5mdCMP + dCMP} \times 100$$

where 5mdCMP and dCMP are expressed as molar equivalents.

3.5. Quantification of Nucleotides Using a Chart Recorder

The method of quantification outlined above is only suitable when computerized data acquisition and analysis are available. Alternatively, if data analysis is by means of a chart recorder, the system must first be calibrated by injecting known amounts of nucleotide standards onto the column and measuring peak heights.

Deoxyribonucleotide standards are available from Sigma. When setting up the calibration, the range of quantities of nucleotide used should span the quantities present in the test samples, and a linear relationship between peak height or area and nucleotide quantity should be demonstrated over the range.

3.5.1 Making Up Solutions with Known Concentrations of dNMPs

1. Dissolve deoxyribonucleotide-5′-monophosphate standards in water and measure the molarities according to Beer's law:

   absorbance at λ_{max} = ε_{max} × path length (cm) × concentration (moles/L)
   ε_{max} (extinction coefficient) for dCMP = 9.3×10^3, 5mdCMP = 11.8×10^3, TMP = 10.2×10^3, dGMP = 13.7×10^3, dAMP = 15.3×10^3 at neutral pH.

   Care must be taken to ensure that the nucleotides are completely dissolved. Their optical densities should not change over time when completely dissolved.

2. Adjust the concentrations of dCMP, TMP, dGMP, and dAMP so that they are approx 7.5 mM, and adjust the concentration of 5mdCMP to approx 0.3 mM.

3. After adjustment, re-measure the concentrations of each standard. Knowing the exact molarity of each solution, pipet exactly 7.50 μmoles of dCMP and dGMP, 10 μmoles of TMP and dAMP, and 0.34 μmoles 5mdCMP into a fresh polypropylene tube (approx 40 μL of each). Care should be taken to ensure that the pipet is properly calibrated before dispensing these amounts, as errors at this stage will result in a failure of nucleotide ratios to balance when the calibration is used to compute molar ratios. Adjust the volume of the mixture to 1 mL by addition of DNA digestion buffer (containing 1 part DNase I digestion buffer to 2 parts nuclease P1 digestion buffer). Dilutions of this deoxyribonucleotide monophosphate mixture can now be used for the calibration.

4. Pipet 100, 90, 80, 70, 60, 50, 40, 30, 20, and 10 μL of this mixture into fresh polypropylene tubes and make up the volume of each to 100 μL with DNA digestion buffer. Sequentially, inject 50 μL of each mixture onto the HPLC column and record the peak areas and peak heights of all nucleotides.

5. Perform a regression analysis of nucleotide quantity against peak height for each nucleotide. In terms of dCMP, this range of dilutions should contain 15 nmols to 1.5 nmoles of nucleotide. The R^2 value for each regression line should be > 0.99 and the regression equation for each nucleotide can be used for converting

any measurement of peak height in a test sample to the corresponding nucleotide molar quantity.

The accuracy of the calibration can be checked by analyzing a segment of DNA (such as φX174 DNA or SV40 DNA) for which the sequence, and therefore predicted nucleotide ratios, is entirely known. Highly purified commercial preparations of these DNAs are available. If the nucleotide ratios found are not as predicted, appropriate adjustments can be made to the calibration. Although this system can be used to verify the accuracy of the calibrations for the major deoxyribonucleotides, it cannot be used for checking the calibration of 5mdCMP, as 5mC is not present in these DNAs. However, the accuracy of the 5mC calibration can be checked indirectly by ensuring that the molar ratio of total dCMP:dGMP is found to be 1 (or close to 1) when analyzing partially methylated DNAs such as human DNA.

3.6. Tests of Reproducibility

If all the points in the regression analysis fit closely to a straight line, this is a good indication that the coefficient of variation is low, but variation here can be due to pipetting errors. All regression lines should pass through (or near to) the origin, and if this is the case, the ratios between nucleotides will be the same over the range of values tested. As a final confirmation, replicate measurements (at least 5) of a test sample should be made to ensure consistency of measurement. The coefficient of variation (standard deviation/mean × 100) should be less than 5% for 5mdCMP/(5mdCMP + dCMP).

4. Notes

The most common problems encountered with this technique are:

1. Incomplete hydrolysis of contaminating RNA. This may lead to baseline variation but if contamination is considerable, it can interfere with the measurement of the dNMP peaks. In **Fig. 3**, contaminating RNA has produced some baseline noise in the sample from wild-type ES cells. This artefact is not present in the sample from DNA methyltransferase-deficient ES cells.
2. Poor performance of the HPLC system leading to a coefficient of variation of >5%. It is best to use a system that is up and running and regularly serviced. For reproducible results, care should be taken to ensure that the system is operating at constant pressure. This is best achieved by connecting the pressure monitor to a chart recorder and measuring the variation in pressure with each pump cycle.

References

1. Gama-Sosa, M. A., Slagel, V. A., Trewyn, R. W., Oxenhandler, R., Kuo, K. C., Gehrke, C. W., and Ehrlich, M. (1983) The 5-methylcytosine content of DNA from human tumours. *Nucleic Acids Res.* **11,** 6883–6894.

2. Kuo, K. C., McCune, R. A., and Gehrke, C. W. (1980) Quantitative reversed-phase high-performance liquid chromatographic determination of major and modified deoxyribonucleosides in DNA. *Nucleic Acids Res.* **8,** 4763–4776.
3. Gehrke, C. W., McCune, R. A., Gama-Sosa, M. A., Ehrlich, M., and Kuo, K. C. (1984) Quantitative reversed-phase high-performance liquid chromatography of major and modified nucleosides in DNA. *J. Chromatogr.* **301,** 199–219.
4. Pfeifer, G. P., Steigerwald, S., Boehm, T. L. J., and Drahovsky, D. (1988) DNA methylation levels in acute human leukaemia. *Cancer Lett.* **39,** 185–192.
5. Eick, D., Fritz, H.-J., and Doerfler, W. (1983) Quantitative determination of 5-methylcytosine in DNA by reverse-phase high-performance liquid chromatography. *Anal. Biochem.* **135,** 165–171.
6. Sinsheimer, R. L. (1954) The action of pancreatic deoxyribonuclease. I. Isolation of mono- and dinucleotides. *J. Biol. Chem.* **208,** 445–459.
7. Lei, H., Oh, S. P., Okano, M., Juttermann, R., Goss, K. A., Jaenisch, R., and Li, E. (1996) De novo DNA cytosine methyltransferase activities in mouse embryonic stem cells. *Development.* **122(10),** 3195–3205.

4

Methylation Analysis by Chemical DNA Sequencing

Piroska E. Szabó, Jeffrey R. Mann, and Gerd P. Pfeifer

1. Introduction

The presence of 5-methylcytosine as a modified base in DNA was discovered many decades ago. Surprisingly, however, and despite intense research efforts, the principal function of DNA methylation is still unknown. The CpG dinucleotide is the predominant if not exclusive target sequence for methylation by mammalian DNA methyltransferases. The analysis of DNA methylation at single-nucleotide resolution (genomic sequencing) has long been considered technically difficult, at least in mammalian cells. Recently, techniques have been developed that give a sufficient specificity and sensitivity for analysis of the methylation of single-copy genes by DNA-sequencing techniques *(1,2)*. Currently, the most widely used method is based on bisulfite-induced deamination of cytosines followed by polymerase chain reaction (PCR) and DNA sequencing *(2)*. Chemical DNA sequencing combined with ligation-mediated PCR (LM-PCR) is an alternative method for determination of genomic methylation patterns *(1)*. LM-PCR is based on the ligation of an oligonucleotide linker onto the 5' end of each DNA molecule that was created by a strand-cleavage reaction during chemical DNA sequencing. This ligation reaction provides a common sequence on all 5' ends allowing exponential PCR to be used for signal amplification. One microgram of mammalian DNA per lane is more than sufficient to obtain good-quality DNA sequence ladders. The general LM-PCR procedure used for methylation analysis by chemical DNA sequencing is outlined in **Fig. 1**. The first step of the procedure is modification and cleavage of DNA with hydrazine and piperidine, generating DNA molecules with a 5'-phosphate group. Hydrazine reacts with cytosines but not 5-methylcytosines. Strand cleavage by piperidine through beta-elimination

From: *Methods in Molecular Biology, vol. 200: DNA Methylation Protocols*
Edited by: K. I. Mills and B. H. Ramsahoye © Humana Press Inc., Totowa, NJ

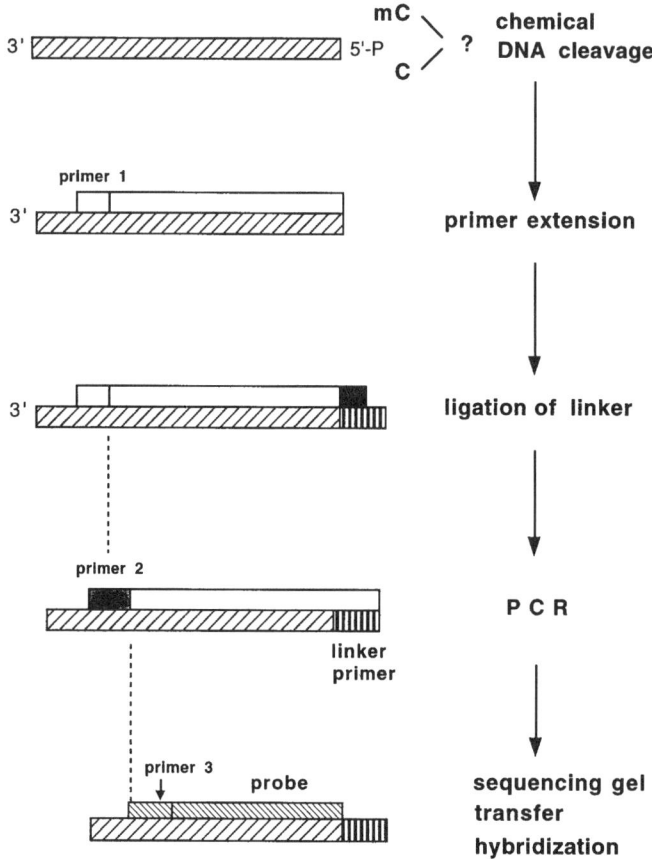

Fig. 1. Outline of the ligation-mediated PCR procedure for analysis of methylation patterns after chemical sequencing. The individual steps include chemical modification and cleavage of DNA, annealing and extension of primer 1, ligation of the linker, PCR amplification of gene-specific fragments with primer 2 and the linker-primer, detection of the sequence ladder by gel electrophoresis, electroblotting, and hybridization with a single-stranded probe.

produces signals at the positions of all cytosines but 5-methylcytosines are recognized by a gap in the sequence ladder. In LM-PCR, primer extension of a gene-specific oligonucleotide (primer 1) generates molecules that have a blunt end on one side. Linkers are ligated to these blunt ends, and then an exponential PCR amplification of the linker-ligated fragments is done using the longer oligonucleotide of the linker (linker-primer) and a second gene-specific primer (primer 2). After 18–20 PCR amplification cycles, the DNA fragments are separated on a sequencing gel, electroblotted onto nylon membranes, and hybridized with a gene-specific probe to visualize the sequence ladders (*1*).

The LM-PCR method has been used for determination of DNA cytosine methylation patterns in various genes *(1,3–10)*. It is possible to use LM-PCR to determine the methylation pattern of restriction sites by a highly sensitive Southern-blot assay that requires only 10 ng of DNA *(11)*. This assay is quantitative and, unlike other PCR-based methylation assays, it gives a simultaneous positive display for methylated and unmethylated sites.

Figure 2 illustrates a methylation analysis obtained by LM-PCR. The sequences shown are from the far upstream region of the maternally expressed, imprinted mouse H19 gene *(9)*. Differential methylation at CpG sites is present between androgenetic, wild-type, and parthenogenetic embryonic stem cells. These differences are clearly apparent when a comparison is made with cloned, unmethylated DNA.

Is there any advantage of using LM-PCR rather than bisulfite sequencing? Published reports that used the bisulfite method often contain data on significant amounts of non-CpG methylation and on asymetrically methylated CpG sites. This has never been observed with LM-PCR. As pointed out by Rein et al. *(12)*, bisulfite sequencing can produce serious artifacts by incomplete denaturation of GC-rich DNA, incomplete deamination of cytosines, or incomplete resistance of 5-methylcytosines. Also, deaminated and nondeaminated sequences may be amplified with different efficiencies in the PCR reaction *(13)*. None of these problems is a concern in LM-PCR. In LM-PCR, partially methylated sites are readily apparent (*see* **Fig. 2**). Quantitation, which should always involve unmethylated, cloned DNA as a control, is best done by comparing the signal at the cytosine of a CpG site with a neighboring cytosine that is not in a CpG. With bisulfite sequencing the quantitative determination of methylation patterns often involves sequencing of a large number of cloned PCR products (e.g., ref. *14*), which can be time-consuming and expensive. However, it has the advantage of obtaining the methylation patterns of single molecules, which cannot be done by LM-PCR. One other significant advantage of LM-PCR is that information about protein binding and chromatin structure can easily be obtained by in vivo footprinting experiments done with the same sets of primers used for determining the methylation pattern *(9,15–18)*.

We suggest that LM-PCR should be used to confirm data obtained with bisulfite sequencing, in particular when artifacts are suspected. LM-PCR is often perceived technically more difficult than bisulfite sequencing. The detailed protocol that follows should alleviate these concerns.

2. Materials

1. Buffer A: 0.3 M sucrose, 60 mM potassium chloride, 15 mM sodium chloride, 60 mM Tris-HCl, pH 8.0, 0.5 mM spermidine, 0.15 mM spermine, 2 mM ethylenediaminetetraacetic acid (EDTA).

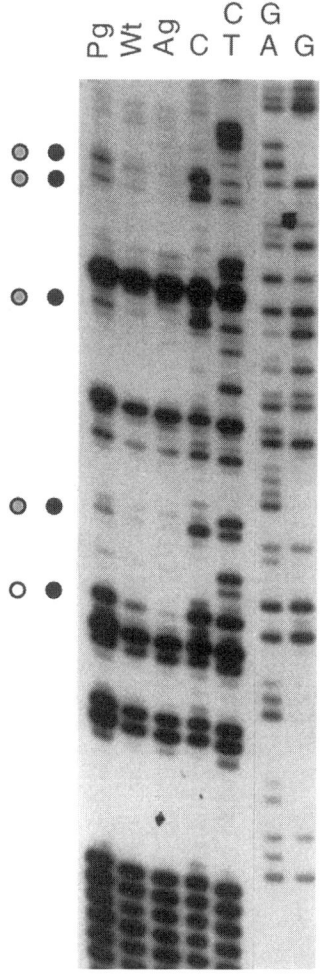

Fig. 2. Detection of methylated cytosines in a mammalian gene. The region analyzed contains sequences 3.9 kilobases upstream of the promoter of the imprinted mouse H19 gene. DNA was obtained from parthenogenetic embryonic stem cells (Pg), wild-type embryonic stem cells (Wt) or androgenetic embryonic stem cells (Ag). The lane labeled C contains unmethylated control DNA obtained from a lambda vector carrying a genomic copy of the H19 gene. The lanes labeled CT, GA, and G are DNAs from wild-type embryonic stem cells subjected to the C+T-, G+A-, and G-specific chemical sequencing reactions. Open, gray, and black circles indicate unmethylated, partially methylated, or fully methylated CpGs in Pg or Ag cells, respectively.

2. Nonidet P40.
3. Buffer B: 150 mM NaCl, 5 mM EDTA, pH 8.0.
4. Buffer C: 20 mM Tris-HCl, pH 8.0, 20 mM NaCl, 20 mM EDTA, 1% sodium dodecyl sulfate (SDS).
5. Proteinase K.
6. DNase-free RNAase A.
7. Phenol: Equilibrate with 0.1 M Tris-HCl, pH 8.0.
8. Chloroform.
9. Ethanol.
10. 3 M sodium acetate, pH 5.2.
11. TE buffer: 10 mM Tris-HCl, pH 7.6, 1 mM EDTA.
12. DMS buffer: 50 mM sodium cacodylate, 1 mM EDTA, pH 8.0.
13. DMS (dimethylsulfate, >99%, Aldrich, Milwaukee, WI). DMS is a highly toxic chemical and should be handled in a well-ventilated hood. DMS waste (including plastic material) is detoxified in 5 M NaOH. DMS is stored under nitrogen at 4°C.
14. DMS stop: 1.5 M sodium acetate, pH 7.0, 1 M 2-mercaptoethanol.
15. Formic acid (Fluka, Ronkonkoma, NY).
16. Hydrazine (anhydrous, Aldrich). Hydrazine is a highly toxic and should be handled in a well-ventilated hood. Hydrazine waste (including plastic material) is detoxified in a solution of 3 M ferric chloride. Hydrazine is stored under nitrogen at 4°C in an explosion-proof refrigerator. The bottle should be replaced at least every 6 mo.
17. Hz-stop: 0.3 M sodium acetate, pH 7.5, 0.1 mM EDTA.
18. Ethanol pre-cooled to –70°C.
19. 3 M sodium acetate, pH 5.2.
20. 75% ethanol.
21. Piperidine (>99%, Fluka), stored under nitrogen at –20°C. The 1 M solution is prepared fresh.
22. Oligonucleotide primers for primer extension: The primer used, as primer 1 (Sequenase primer) is a 15- to 20-mer with a calculated Tm of 48°C to 56°C (*see* **Note 1**). Prepare primers as stock solutions of 50 pmoles/µL in TE buffer and keep at –20°C.
23. 5X Sequenase buffer: 250 mM NaCl, 200 mM Tris-HCl, pH 7.7.
24. Mg-DTT-dNTP mix: 20 mM MgCl$_2$, 20 mM DTT, 0.25 mM of each dNTP.
25. Sequenase 2.0 (USB, Cleveland, OH): 13 units/µL.
26. 300 mM Tris-HCl, pH 7.7.
27. 2 M Tris-HCl, pH 7.7.
28. Linker: Prepare the double-stranded linker by annealing a 25-mer (5′-GCGGT GACCCGGGAGATCTGAATTC, 20 pmoles/µL) to an 11-mer (5′-GAATTCA GATC, 20 pmoles/µL) in 250 mM Tris-HCl, pH 7.7, by heating to 95°C for 3 min and gradually cooling to 4°C over a time period of 3 h. Linkers can be stored at –20°C for at least 3 mo. They are thawed and kept on ice.
29. Ligation mix: 13.33 mM MgCl$_2$, 30 mM DTT, 1.66 mM ATP, 83 µg/mL bovine serum albumin (BSA), 3 units/reaction T4 DNA ligase (Promega, Madison, WI), and 100 pmoles linker/reaction (= 5 µL linker).

30. *Escherichia coli* tRNA.
31. 2X Taq polymerase mix: 20 mM Tris-HCl, pH 8.9, 80 mM NaCl, 0.02% gelatin, 4 mM MgCl$_2$, and dNTPs at 0.4 mM each.
32. Oligonucleotide primers for PCR: The primer used in the amplification step (primer 2) is a 20- to 30-mer with a calculated Tm between 60 and 68°C (*see* **Note 2**). 10 pmoles of the gene-specific primer (primer 2) and 10 pmoles of the 25-mer linker-primer (5′-GCGGTGACCCGGGAGATCTGAATTC) are used per reaction along with 3 units Taq polymerase, and these components can be included in the 2X Taq polymerase mix.
33. Taq polymerase
34. Mineral oil.
35. 400 mM EDTA, pH 7.7.
36. Formamide loading buffer: 94% formamide, 2 mM EDTA, pH 7.7, 0.05% xylene cyanol, 0.05% bromophenol blue.
37. Acrylamide.
38. Bisacrylamide.
39. Urea.
40. 1 M TBE: 1 M Tris, 0.83 M boric acid, 10 mM EDTA, pH 8.3.
41. Whatman 3MM and Whatman 17 paper.
42. Nylon membranes.
43. Electroblotting apparatus (Owl Scientific, Cambridge, MA).
44. An appropriate plasmid or PCR product containing the sequences of interest.
45. Oligonucleotide primer to make the hybridization probe (primer 3). This primer is used together with the cloned template and Taq polymerase to make single-stranded hybridization probes (*see* **Note 3**).
46. 32P-dCTP(3000 Ci/mmol)
47. 7.5 M Ammonium acetate.
48. Hybridization buffer: 0.25 M sodium phosphate, pH 7.2, 1 mM EDTA, 7% SDS, 1% BSA.
49. Washing buffer: 20 mM sodium phosphate, pH 7.2, 1 mM EDTA, 1% SDS.
50. Kodak XAR-5 film.

3. Methods

3.1. DNA Isolation

1. Isolate nuclei by adding 10 mL of buffer A containing 0.5% Nonidet P40 to the cells. This step will release the nuclei and removes most of the cytoplasmic RNA. Transfer the suspension to a 50-mL tube. Incubate on ice for 5 min.
2. Centrifuge at 1,000g for 5 min at 4°C.
3. Wash the nuclear pellet once with 15 mL of buffer A.
4. Re-suspend nuclei thoroughly in 2–5 mL of buffer B, add one volume of buffer C, containing 600 µg/mL of proteinase K (added just before use). Incubate for 2 h at 37°C.

5. Add DNase-free RNAase A to a final concentration of 100 µg/mL. Incubate for 1 h at 37°C.
6. Extract with one volume of buffer-saturated phenol. Then, extract with 0.5 volumes of phenol and 0.5 volumes of chloroform. Repeat this step until the aqueous phase is clear and no interface remains. Finally, extract with 1 volume of chloroform.
7. Add 0.1 volumes of 3 *M* sodium acetate, pH 5.2, and precipitate the DNA with 2.5 volumes of ethanol at room temperature.
8. Centrifuge at 2,000*g* for 1 min. Wash the pellet with 75% ethanol and air-dry briefly.
9. Dissolve the DNA in TE buffer to a concentration of approx 0.2 µg/mL. Keep at 4°C overnight.

3.2. Chemical DNA Sequencing

Genomic DNA is chemically sequenced by the Maxam-Gilbert procedure *(19)*. The conditions below work well for 40–80 µg of DNA. DNA may be digested with a restriction enzyme to reduce viscosity. Carrier *E. coli* DNA can be added if the amount of DNA is insufficient. If necessary, the DNA solutions can be concentrated in a vacuum concentrator or the DNA can be ethanol-precipitated before chemical sequencing. In the analysis of methylation patterns, an unmethylated control DNA is required. This can be a plasmid, lambda DNA, or a PCR product. It is necessary to dilute the cloned DNA with carrier DNA down to the level of a single-copy gene. This involves a step-wise dilution of at least one hundred thousand-fold. If desired, a methylated control DNA can be obtained from this cloned DNA by incubation with the CpG-specific prokaryotic DNA methyltransferase MSssI under conditions described elsewhere *(20)*. Because methylation analysis involves other parallel base-specific sequencing reactions, we describe all reactions needed to read the complete sequence.

3.2.1. G-Reaction

1. Mix carefully on ice: 5 µL genomic DNA (40–80 µg, 200 µL DMS buffer, 1 µL DMS.
2. Incubate at 20°C for 3 min.
3. Add 50 µL DMS stop.
4. Add 750 µL pre-cooled ethanol (–70°C).

3.2.2. G+A Reaction

1. Mix on ice: 11 µL genomic DNA (40–80 µg), 25 µL formic acid.
2. Incubate at 20°C for 10 min.
3. Add 200 µL DMS stop.
4. Add 750 µL pre-cooled ethanol (–70°C).

3.2.3. T+C Reaction

1. Mix well on ice: 20 µL genomic DNA (40–80 µg), 30 µL hydrazine.
2. Incubate at 20°C for 15 min.
3. Add 200 µL Hz-stop.
4. Add 750 µL pre-cooled ethanol (–70°C).

3.2.4. C Reaction

1. Mix well on ice: 5 µL genomic DNA (40–80 µg), 15 µL 5 M NaCl. Mix well, then add 30 µL hydrazine.
2. Incubate at 20°C for 15 min.
3. Add 200 µL Hz-stop.
4. Add 750 µL pre-cooled ethanol (–70°C).

Process all samples as follows:

5. Keep samples in a dry ice/ethanol bath for 15 min.
6. Spin 15 min at 14,000 rpm in an Eppendorf centrifuge at 0–4°C.
7. Take out supernatant, re-spin.
8. Resuspend pellet in 225 µL water.
9. Add 25 µL 3 M sodium acetate, pH 5.2.
10. Add 750 µL pre-cooled ethanol (–70°C).
11. Put on dry ice, 15 min.
12. Spin 10 min at 14,000 rpm in Eppendorf centrifuge at 0–4°C.
13. Take out supernatant, re-spin.
14. Wash with 1 mL 75% ethanol; spin 5 min in Eppendorf centrifuge.
15. Dry pellet in a Speedvac concentrator.
16. Dissolve pellet in 100 µL 1 M piperidine (freshly diluted).
17. Secure caps with Teflon tape and lid locks.
18. Heat at 90°C for 30 min in a heat block.
19. Transfer to a new tube.
20. Add 1/10 vol. 3 M sodium acetate, pH 5.2.
21. Add 2.5 vols Ethanol.
22. Put on dry ice, 20 min.
23. Spin 15 min at 14,000 rpm in Eppendorf centrifuge at 0–4°C.
24. Wash twice with 75% ethanol.
25. Remove traces of remaining piperidine by drying the sample overnight in a Speedvac concentrator. Dissolve DNA in water to a concentration of about 0.5 µg/µL.
26. Determine the cleavage efficiency by running 1 µg of the samples on a 1.5% agarose gel. Most fragments should be in a size range between 50 and 800 nucleotides (*see* **Note 4**).

3.3. Ligation-Mediated PCR

1. Mix in a siliconized 1.5 mL tube: 1 µg of cleaved DNA, 0.6 pmoles of primer 1, and 3 µL of 5X Sequenase buffer in a final volume of 15 µL.

Chemical DNA Sequencing

2. Incubate at 95°C for 3 min, then at 45°C for 30 min.
3. Cool on ice, spin 5 s.
4. Add 7.5 µL cold, freshly prepared Mg-DTT-dNTP mix.
5. Add 1.5 µL Sequenase, diluted 1:4 in cold 10 mM Tris, pH 7.7.
6. Incubate at 48°C, 10 min, then cool on ice (*see* **Note 5**).
7. Add 6 µL 300 mM Tris-HCl, pH 7.7.
8. Incubate at 67°C, 15 min (heat inactivation of Sequenase).
9. Cool on ice, spin 5 s.
10. Add 45 µL of freshly prepared ligation mix.
11. Incubate overnight at 18°C.
12. Incubate 10 min at 70°C (heat inactivation of ligase).
13. Add 8.4 µL 3 M sodium acetate, pH 5.2, 10 µg *E. coli* tRNA, and 220 µL ethanol.
14. Put samples on dry ice for 15 min.
15. Centrifuge 10 min at 4°C in an Eppendorf centrifuge.
16. Wash pellets with 950 µL 75% ethanol.
17. Remove ethanol residues in a SpeedVac.
18. Dissolve pellets in 50 µL H$_2$0 and transfer to 0.5 mL siliconized tubes.
19. Add 50 µL freshly prepared 2X Taq polymerase mix containing the primers and the enzyme and mix by pipetting.
20. Cover samples with 50 µL mineral oil and spin briefly.
21. Cycle 18–20 times at 95°C, 1 min, 60–66°C, 2 min, and 76°C, 3 min.
    Add 1 unit of fresh Taq polymerase per sample together with 10 µL reaction buffer. Incubate 10 min at 74°C (*see* **Note 6**).
23. Add sodium acetate to 300 mM and EDTA to 10 mM to stop reaction and 10 µg tRNA.
24. Extract with 70 µL of phenol and 120 µL chloroform (premixed).
25. Add 2.5 vol. of ethanol and put on dry ice for 15 min.
26. Centrifuge samples 10 min in an Eppendorf centrifuge at 4°C.
27. Wash pellets in 1 mL 75% ethanol.
28. Dry pellets in Speedvac.

3.4. Sequencing Gel Analysis of Reaction Products

1. Prepare a 8% polyacrylamide gel (acrylamide:bisacrylamide = 29:1) containing 7 M urea and 0.1 M TBE, 0.4 mm thick and approx 60 cm long.
2. Dissolve pellets in 1.5 µL of water and add 3 µL formamide loading buffer.
3. Heat samples to 95°C for 2 min prior to loading.
4. Load only one half of the sample or less using a very thin flat tip.
5. Run the gel until the xylene cyanol marker reaches the bottom. Fragments below the xylene cyanol dye do not hybridize significantly.
6. After the run, transfer the gel (i.e., the bottom 40 cm of it) to Whatman 3MM paper and cover with Saran Wrap.
7. Electroblotting of the gel piece is performed with a transfer box available from Owl Scientific (*see* **Note 7**). Pile three layers of Whatman 17 paper, 43 × 19 cm, presoaked in 90 mM TBE, onto the lower electrode. Squeeze the paper with a

roller to remove air bubbles between the paper layers. Place the gel piece covered with Saran wrap onto the paper and remove the air bubbles between the gel and the paper by wiping over the Saran wrap with a soft tissue. Remove the plastic wrap and cover the gel with a nylon membrane cut somewhat larger than the gel and presoaked in 90 mM TBE. Put three layers of presoaked Whatman 17 paper onto the nylon membrane carefully removing trapped air with a glass rod. Place the upper electrode onto the paper. The electroblotting procedure is performed at a current of 1.6 A. After 30 min, the nylon membrane is removed and the DNA side is marked. A high ampere power supply is required for this transfer. After electroblotting, dry the membrane briefly at room temperature. Then cross-link the DNA by UV irradiation in a commercially available crosslinker.

8. Soak the nylon membranes in 50 mM TBE buffer, roll them onto a pipet and transfer to 250-mL plastic or glass hybridization oven cylinders so that the membranes stick completely to the walls of the cylinders without air pockets.
9. Pre-hybridize the membrane with 15 mL hybridization buffer for 10 min at 62°C.
10. To prepare single-stranded probes for hybridization, which are 200–300 nucleotides long, use repeated primer extension by Taq polymerase with a single primer (primer 3) on a double-stranded template DNA *(21)*. This can be either plasmid DNA restriction-cut approx 200–300 nucleotides 3′ to the binding site of primer 3 or a PCR product that contains the target area of interest. Mix 50 ng of the restriction-cut plasmid DNA (or 10 ng of the gel-purified PCR product) with primer 3 (20 pmoles), 100 µCi of [^{32}P]dCTP, 10 µM of the other three dNTPs, 10 mM Tris-HCl, pH 8.9, 40 mM NaCl, 0.01% gelatin, 2 mM MgCl$_2$, and 3 units of Taq polymerase in a volume of 100 µL. Perform 35 cycles at 95°C (1 min), 60–66°C (1 min), and 75°C (2 min). Recover the probe by phenol/chloroform extraction, addition of ammonium acetate to a concentration of 0.7 M, ethanol precipitation at room temperature, and centrifugation.
11. Dilute the labeled probe into 5 mL hybridization buffer and hybridize for 16 h at 62°C.
12. Following hybridization, wash each nylon membrane with 2 L of washing buffer at 60°C. Perform several washing steps in a dish at room temperature with prewarmed buffer. After washing, dry the membrane briefly at room temperature, cover with Saran wrap and expose to Kodak XAR-5 films. If the procedure has been done without error, a result can be seen after 0.5–8 h of exposure with intensifying screens at –80°C.

4. Notes

1. Calculation of the Tm is done with a computer program *(22)*. Primers do not need to be gel-purified, if the oligonucleotide synthesis quality is sufficiently good (less than 5% of n–1 or n+1 material on analytical polyacrylamide gels). If a very specific target area is to be analyzed, primer 1 should be located approx 100 nucleotides upstream of this target.

2. Primer 2 is designed to extend 3' to primer 1. Primer 2 can overlap several bases with primer 1, but we have also had good results with a second primer that overlapped only one or two bases with the first.
3. Primer 3 is the primer used to make the single-stranded hybridization probe. It should be on the same strand just 3' to the amplification primer (primer 2) and should have a Tm of 60–68°C. It should not overlap more than 8–10 bases with primer 2.
4. It is important that the average fragment sizes are similar for samples that need to be compared directly. The approximate amount of DNA used in the LM-PCR reactions can be estimated from the relative amount of DNA visible on the gels. This estimation allows one to obtain similar band intensities on the sequencing gel in all lanes without having to re-run the sequencing gel to achieve equal loading.
5. At 48°C, the terminal transferase activity of Sequenase is reduced resulting in more blunt end formation. If sequence ladders are still incomplete, which may be the case in particularly purine-rich areas, Vent exo- may be used together with the manufacturer's recommended buffer as described elsewhere *(23)*. In this case, the DNA is precipitated before the ligation step.
6. This step is to completely extend all DNA fragments and add an extra nucleotide through Taq polymerase's terminal transferase activity. If this step is omitted, double bands may occur.
7. The advantages of the hybridization approach over the end-labeling technique *(24)* have been discussed previously *(15)*.

Acknowledgment

National Institute of Environmetal Health Sciences grant (ES06070) to G.P.P. supported this work.

References

1. Pfeifer, G. P., Steigerwald, S. D., Mueller, P. R., Wold, B., and Riggs, A. D. (1989) Genomic sequencing and methylation analysis by ligation mediated PCR. *Science* **246,** 810–813.
2. Frommer, M., McDonald, M. E., Millar, D. S., Collis, C. M., Watt, F., Grigg, G. W., et al. (1992) A genomic sequencing protocol that yields a positive display of 5-methylcytosine residues in individual DNA strands. *Proc. Natl. Acad. Sci. USA* **89,** 1827–1831.
3. Pfeifer, G. P., Tanguay, R. L., Steigerwald, S. D., and Riggs, A. D. (1990) In vivo footprint and methylation analysis by PCR-aided genomic sequencing: comparison of active and inactive X chromosomal DNA at the CpG island and promoter of human PGK-1. *Genes Dev.* **4,** 1277–1287.
4. Pfeifer, G. P., Steigerwald, S. D., Hansen, R. S., Gartler, S. M., and Riggs, A. D. (1990) Polymerase chain reaction-aided genomic sequencing of an X chromosome-linked CpG island: methylation patterns suggest clonal inheritance,

CpG site autonomy, and an explanation of activity state stability. *Proc. Natl. Acad. Sci. USA* **87,** 8252–8256.
5. Rideout III, W. M., Coetzee, G. A., Olumi, A. F., and Jones, P. A. (1990) 5-Methylcytosine as an endogenous mutagen in the human LDL receptor and p53 genes. *Science* **249,** 1288–1290.
6. Reddy, P. M. S., Stamatoyannopoulos, G., Papayannopoulou, T., and Shen, C. K. J. (1994) Genomic footprinting and sequencing of human β-globin locus. *J. Biol. Chem.* **269,** 8287–8295.
7. Hornstra, I. K. and Yang, T. P. (1994) High-resolution methylation analysis of the human hypoxanthine phosphoribosyltransferase gene 5′ region on the active and inactive X chromosomes: correlation with binding sites for transcription factors. *Mol. Cell. Biol.* **14,** 1419–1430.
8. Tornaletti, S. and Pfeifer, G. P. (1995) Complete and tissue-independent methylation of CpG sites in the p53 gene: implications for mutations in human cancers. *Oncogene* **10,** 1493–1499.
9. Szabó, P. E., Pfeifer, G. P., and Mann, J. R. (1998) Characterization of novel parental-specific epigenetic modifications upstream of the imprinted mouse H19 gene. *Mol. Cell. Biol.* **18,** 6767–6776.
10. You, Y. H., Halangoda, A., Buettner, A., Hill, K., Sommer, S., and Pfeifer, G. P. (1998) Methylation of CpG dinucleotides in the lacI gene of the Big Blue' transgenic mouse. *Mutation Res.* **420,** 55–65.
11. Steigerwald, S. D., Pfeifer, G. P., and Riggs, A. D. (1990) Ligation- mediated PCR improves the sensitivity of methylation analysis by restriction enzymes and detection of specific DNA strand breaks. *Nucleic Acids Res.* **18,** 1435–1439.
12. Rein, T., DePamphelis, M. L., and Zorbas, H. (1998) Identifying 5-methylcytosine and related modifications in DNA genomes. *Nucleic Acids Res.* **26,** 2255–2264.
13. Warnecke, P. M., Stirzaker, C., Melki, J. R., Millar, D. S., Paul, C. L., and Clark, S. J. (1997) Detection and measurement of PCR bias in quantitative methylation analysis of bisulphite-treated DNA. *Nucleic Acids Res.* **25,** 4422–4426.
14. Stirzaker, C., Millar, D. S., Paul, C. L., Warnecke, P. M., Harrison, J., Vincent, P., et al. (1997) Extensive DNA methylation spanning the Rb promoter in retinoblastoma tumors. *Cancer Res.* **57,** 2229–2237.
15. Pfeifer, G. P. and Riggs, A. D. (1993) Genomic footprinting by ligation mediated polymerase chain reaction, In: *Methods in Molecular Biology*, vol. 15, (White, B.A., ed.), PCR Protocols: Current Methods and Applications, Humana Press, Totowa, NJ, pp. 153–168.
16. Tommasi, S. and Pfeifer, G. P. (1995) In vivo structure of the human cdc2 promoter: release of a p130/E2F-4 complex from sequences immediately upstream of the transcription initiation site coincides with induction of cdc2 expression. *Mol. Cell. Biol.* **15,** 6901–6913.
17. Tornaletti, S. and Pfeifer, G. P. (1995) UV-light as a footprinting agent: modulation of UV-induced DNA damage by transcription factors bound at the promoters of three human genes. *J. Mol. Biol.* **249,** 714–728.

18. Chin, P. L., Momand, J., and Pfeifer, G. P. (1997) In vivo evidence for binding of p53 to consensus binding sites in the p21 and GADD45 genes in response to ionizing radiation. *Oncogene* **15,** 87–99.
19. Maxam, A. M. and Gilbert, W. (1980) Sequencing end-labeled DNA with base-specific chemical cleavages. *Methods Enzymol.* **65,** 499–560.
20. Denissenko, M. F., Chen, J. X., Tang, M. S., and Pfeifer, G. P. (1997) Cytosine methylation determines hot spots of DNA damage in the human P53 gene. *Proc. Natl. Acad. Sci. USA* **94,** 3893–3898.
21. Törmänen, V. T. and Pfeifer, G. P. (1992) Mapping of UV photoproducts within ras protooncogenes in UV-irradiated cells: correlation with mutations in human skin cancer. *Oncogene* **7,** 1729–1736.
22. Rychlik, W. and Rhoads, R. E. (1989) A computer program for choosing optimal oligonucleotides for filter hybridization, sequencing and in vitro amplification of DNA. *Nucleic Acids Res.* **17,** 8543–8551.
23. Komura, J.-I. and Riggs, A. D. (1998) Terminal transferase dependent PCR: a versatile and sensitive method for in vivo footprinting and detection of DNA adducts. *Nucleic Acids Res.* **26,** 1807–1811.
24. Mueller, P. R. and Wold, B. (1989) In vivo footprinting of a muscle specific enhancer by ligation mediated PCR. *Science* 246, 780–786.

5

Methylation-Sensitive Restriction Fingerprinting

Catherine S. Davies

1. Introduction

Methylation of cytosine residues is an almost ubiquitous finding of higher organisms *(1)*. The majority of this methylation occurs at the dinucleotide CpG (where p denotes a phosphate group) *(2)*. CpG sites are distributed throughout the genome with clusters of the sequence being found in the 5' promoter region of housekeeping genes, in groups known as CpG islands. These short stretches of DNA have a have a C and G base composition, which in mammals and avians is estimated to be 10 times higher than in bulk DNA *(3)*.

Typically, CpG islands are unmethylated, the notable exception being the inactive X-chromosome *(4)*. *De novo* methylation of the island is prevented by the Sp1 elements, which are located upstream of the gene *(5)*. The maintenance of the unmethylated state is necessary for the expression of the related gene *(6)*.

Investigations in recent years have identified a number of alterations in DNA methylation common to most transformed cells, which have profound effects on DNA structure and function *(6)*. Increased levels of DNA methyltransferase activity and development of localized regions of hypermethylation have been linked to the inactivation of tumor-supressor genes *(7–10)*. This, together with the finding that tumor genomes are susceptible to the loss of methylation from normally methylated sites, suggests that methylation-mediated deregulation of tumor genomes may play a pivotal role in the development and progression of the neoplastic state

As a result, numerous techniques to investigate DNA methylation have been described. The advent of the technique methylation-sensitive restriction fingerprinting (MSRF), has provided a tool to allow the methylation status of CpG sites throughout the entire DNA genome to be analyzed. MSRF is a

From: *Methods in Molecular Biology, vol. 200: DNA Methylation Protocols*
Edited by: K. I. Mills and B. H. Ramsahoye © Humana Press Inc., Totowa, NJ

PCR-based technique that, by virtue of the properties of methylation-sensitive restriction enzymes, is biased towards the study of CpG sites. The DNA fingerprints generated through this procedure are be analyzed and regions of aberrant methylation in tumor genomes relative to normal DNA can be detected.

MSRF has previously been used to study the methylation status of CpG sites in samples derived from patients with chronic myeloid leukemia, leukemia cell lines, and in the study of breast carcinomas *(11)*. In this chapter, the application of the technique MSRF to the detection and sequence characterization of regions of genomic DNA that undergo methylation changes during carcinogenesis is described.

2. Materials
2.1. DNA Extraction

1. 10X Cell lysis buffer: 3 M NH$_4$Cl, 0.1 mM ethylenediaminetetraacetic acid (EDTA), 0.2 M K$_2$CO$_3$.
2. 20% sodium dodecyl sulfate (SDS).
3. 1X PBS: 80 mM Na$_2$HPO$_4$, 20 mM NaH$_2$PO$_4$.2H$_2$0, 10 mM NaCl, pH 7.5.
4. Proteinase K (10 mg/mL) (Gibco BRL).
5. Phenol/Chloroform (50:50 v/v).
6. 100% Ethanol.
7. 70% Ethanol.
8. 1X TE buffer: Tris-HCl, pH 7.5, 0.05 mM EDTA, pH 8.0.

2.2. Restriction Enzyme Digests

1. BstUI (New England Biolabs).
2. MseI (New England Biolabs).
3. BSA (10 mg/mL) (New England Biolabs).
4. Restriction enzyme buffer: 50 mM Tris-HCl, pH 8.0, 10 mM MgCl$_2$.

2.3. PCR Amplification

1. α-^{32}P dCTP (specific activity 3000 mmole/µCi) (Amersham Pharmacia).
2. PCR primers.
   A1   5' - AGCGGCCGCG
   A2   5' - ACCCCAGCCG
   A3   5' - TGGTCGGCGC
   A4   5' - GCACCCGACG
3. 10 mM dNTPs (dATP, dCTP, dGTP, dTTP) (Perkin Elmer).
4. Taq DNA polymerase (Perkin Elmer).
5. Dimethylsulphoxide (DMSO).
6. 10 mM MgCl$_2$.

7. 10X polymerase chain reaction (PCR) Buffer: 100 mM Tris-HCl, pH 8.3, 500 mM KCl, 15 mM MgCl$_2$, 0.01% (w/v) gelatin (Perkin Elmer).
8. Mineral oil.

2.4. Polyacrylamide Gel Electrophoresis

1. Acrylamide (40%, 19:1 Acrylamide:bisacrylamide)
2. 10X TBE: 450 mM Tris-borate, 10 mM EDTA, pH 8.0.
3. N,N,N′,N′-tetraethylenediamine (TEMED).
4. 10% Ammoniumpersulphate.
5. 6X Bromophenol blue loading dye: 0.25% bromophenol blue, 0.25% xylene cyanol, 30% (v/v) glycerol.
6. Kodak X-Omat film.

2.5. Elution of DNA

1. 3 M Sodium acetate, pH 5.2.
2. Glycogen (10 mg/mL) (Sigma).
3. 100% Ethanol.
4. 85% Ethanol.

2.6. Southern Blotting

1. Agarose (electrophoresis grade) (Gibco BRL).
2. 10X TBE.
3. Ethidium bromide (10 mg/mL).
4. 0.125 M HCl.
5. Denaturation solution; 0.66 M NaCl, 0.5 M NaOH.
6. Neutralization solution: 0.66 M NaCl, 0.5 M Tris-HCl.
7. Hybond N nitrocellulose membrane (Amersham Pharmacia).
8. α-^{32}P dCTP: (specific activity 3000 mmole/μCi).
9. Denhart's solution: 0.1 M Ficoll 400, 0.1 M polyvinyl pyrrolidone, 10 mg/mL bovine serum albumin (BSA) portion V (Sigma).
10. Hybridization solution: 1X SSPE, 1% SDS, 1.25X Denhart's solution.
11. 10X SSPE: 3 M NaCl, 0.2 M Na$_2$HPO$_4$, 0.2 M EDTA.
12. Kodak X-Omat film.

2.7. Cloning

1. PCR Trap cloning kit (Genhunter)
2. Luria-Bertani Medium: 10 g/L Bacto-Tryptone, 10 g/L NaCl, 5 g/L Yeast extract.
3. Luria-Bertani Agar: 10 g/L Bacto-Tryptone, 10 g/L NaCl, 5 g/L Yeast extract, 15 g/L Agar.
4. Tetracycline (200 mg/mL) (Sigma).

2.8. Sequencing

1. ABI Big Dye cycle sequencing kit (Perkin Elmer).
2. 3 M NaCl.
3. 95% Ethanol.
4. 70% Ethanol.
5. Sequencing loading dye: 85% de-ionized formamide, 15% 25 mM EDTA with dextran blue (50 mg/mL).

3. Methods

3.1. DNA Isolation

Genomic DNA can be easily isolated from uncoagulated peripheral blood samples. For our purposes 4 mL specimens were collected into vacutainers containing lithium heparin (Becton Dickinson). Genomic DNA can be isolated as detailed below.

1. Erythrocytes are lysed in 2X volume of cell lysis buffer by continuous rotation for 1 h at room temperature.
2. The lysed blood cells are harvested by centrifuged at 1400g for 20 min and the resulting pellet resuspended in 15 mL of 0.2 M Sodium Acetate, 1 mL of 20% SDS and 100 µL of proteinase K. The sample is incubated overnight at 37°C.
3. DNA is extracted by the addition of an equal volume of phenol/chloroform and centrifugation at 3000 rpm for 20 min.
4. The uppermost aqueous phase is collected and the DNA precipitated using 2 volumes of 100% ethanol.
5. The DNA is removed from the solution using a glass hook, washed in 70% ethanol, and dissolved in an appropriate volume of 1X TE buffer to give a final concentration of ~1 µg/µL.

3.2. DNA Digests

The use of methylation-sensitive restriction enzymes allows for discrimination between methylated and unmethylated sequences of the genome. For this application we have used the enzyme *Bst*UI, which will cleave unmethylated CpG sites but leave methylated CpG sequences intact.

1. For each DNA sample, duplicate eppendorf tubes containing 1 µg of DNA are set up.
2. To one of each pair of tubes 10 U of *Bst*UI is added.
3. 10X reaction buffer is added to all tubes and the total volume made up to 10 µL with distilled H$_2$0. All samples are incubated at 60°C for 2 h.
4. Upon completion of the digest reaction the samples are cooled on ice and 10 U of *Mse*I added to both sample tubes (*Mse*I cuts DNA, at non-CpG sites, generating fragments of a size that can be more easily analyzed). Appropriate quantities

of BSA are added to each sample to give a final concentration of 1 µg/µL. The digestion is carried out by incubation at 37°C for 16–20 h.

3.3. PCR Amplification

1. 100 ng of digested DNA is mixed with 0.4 µM of primer A1 and 0.4 µM of either A2, A3, or A4, 5% (v/v) DMSO, 200 µM dNTPs, 12.5 µL α^{32}P dCTP, 0.2U Taq DNA polymerase, 1 mM MgCl$_2$. The sample volume is made up to 10 µL with distilled H$_2$O and overlaid with mineral oil.
2. The DNA is amplified in a Perkin Elmer thermal cycler under the following conditions:
   TD   94°C, 2 min
   TA   40°C, 1 min
   TE   72°C, 2 min × 35 cycles
   Following completion of the final cycle an additional extension period of 10 min at 72°C is added.
3. Samples are best analyzed immediately, however, can be kept overnight at 4°C. If unavoidable, long-term storage should be at –20°C.

3.4. Polyacrylamide Gel Electrophoresis

Polyacrylamide gel electrophoresis (PAGE) can be used to separate small DNA fragments with high resolution

1. An appropriate volume of acrylamide stock (40% 19:1 acrylamide:bis-acrylamide), to produce 4.5–6% gels, is mixed with 1X TBE in a total volume of 100mls.
2. The acrylamide solution is polymerized by the addition of 100 µL of TEMED and 500 µL of 10% APS. The gel is poured between two glass plates (dimensions 52 × 38cm) and allowed to set for 1 h.
3. When polymerized, the gel is placed into the electrophoresis apparatus and the reservoirs filled with 1X TBE buffer.
4. To each sample, 2 µL of bromophenol blue loading dye is added.
5. 4 µL of sample is loaded, in triplicate, onto the gel.
6. The samples are separated by electrophoresis at 1600V/60W for 2–5 h.
7. Following electrophoresis, the glass plates are separated and the gel taken up on to 3MM filter paper. The gel is covered with Saran wrap and placed on a gel dryer at 70°C for 50 min.
8. Exposure of the gel to Kodak X-Omat film for 12–48 h at room temperature allows for restriction fingerprint patterns to be generated.

3.5. Elution of DNA from Polyacrylamide Gels

DNA fragments of interest, identified by analysis of restriction fingerprint profiles (*see* **Subheading 4.**), can be excised from the gel to allow for characterization of the DNA sequence.

1. The area of gel corresponding to the region of interest is spliced out of the restriction fingerprint profile and placed into 0.65 mL Eppendorf tube containing 100 µL of distilled water.
2. Following incubation at room temperature for approx 1 h, the sample is heated to 95°C for 10 min before centrifugation at 800g for 5 min.
3. DNA is precipitated from the resulting supernatant by the addition of 10 µL 3 M sodium acetate, 5 µL glycogen, and 450 µL 100% EtOH.
4. The DNA is harvested by centrifugation at 800g for 10 min and the resulting pellet washed in 250 µL of chilled 85% EtOH.
5. The DNA is resuspended in 10 µL of distilled H_2O.

3.6. Characterization of Abnormally Methylated DNA Fragment

1. Four µL of DNA eluted from the gel fragment is re-amplified in a PCR reaction employing the identical conditions to those described in **Subheading 3.3.**
2. To facilitate the identification of the abnormally methylated DNA fragment a sequencing reaction is performed. The sequence of the PCR product can then be determined. In our applications this was undertaken using the ABI Big Dye cycle sequencing kit in accordance with the manufacturer instructions.

3.7. Confirmation of Altered Methylation Status

Southern hybridization is used to confirm the altered methylation status of the previously characterized DNA fragments. The procedure employed is as follows:

1. Five µg of duplicate genomic DNA samples are digested with 10 U *Msp*I in the presence or absence of 10 U of the methylation-sensitive restriction enzyme *Hpa*II.
2. The digested DNA samples are mixed with 6X loading buffer and separated by electrophoresis, which is carried out at 50 V/mA through a 1% agarose gel in 1X TBE running buffer.
3. Once the bromophenol blue dye front has reached the end of the gel, the electrophoresis is complete. The gel is soaked in 200 mL of neutralizing solution for 10 min, rinsed in distilled water before incubation in denaturation solution for 30 min.
4. The gel is then inverted, placed onto a solid support and the DNA transferred onto nitrocellulose membrane by capillary blotting in a system using 10X SSC as the transfer buffer. This procedure is carried out in accordance with the protocol first described by Southern *(12)*.
5. Transfer of the DNA from the gel to the nitrocelluose membrane takes 12–16 h. When transfer is complete, the DNA is fixed to the nitrocellulose membrane by baking at 80°C for 2 h.
6. Ten µg of cDNA obtained in **Subheading 3.5.** is used as a probe to confirm the altered methylation status. The probe is radiolabeled using a rapid multiprime DNA labeling kit used in accordance with the manufacturers' instructions.

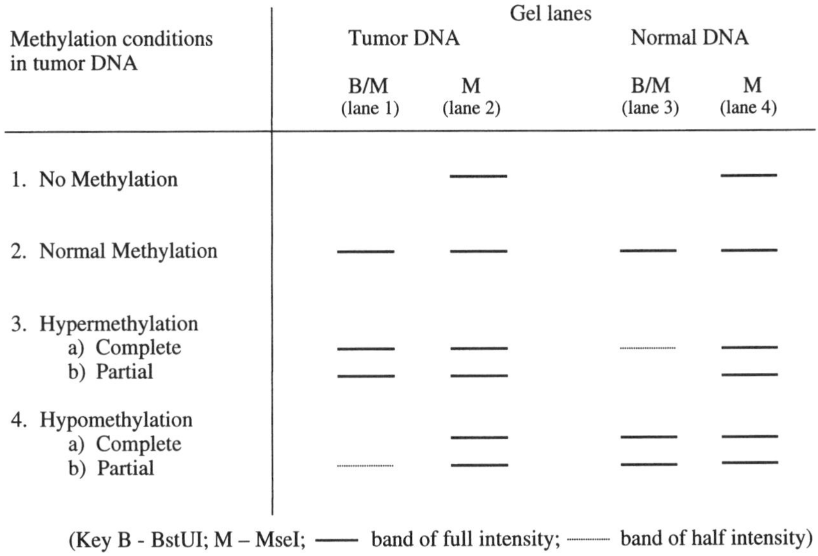

Methylation conditions in tumor DNA	Gel lanes			
	Tumor DNA		Normal DNA	
	B/M (lane 1)	M (lane 2)	B/M (lane 3)	M (lane 4)
1. No Methylation		—		—
2. Normal Methylation	—	—	—	—
3. Hypermethylation a) Complete b) Partial	— —	— —	·······	— —
4. Hypomethylation a) Complete b) Partial	·······	— —	— —	— —

(Key B - BstUI; M – MseI; ——— band of full intensity; ······· band of half intensity)

Fig. 1. Possible banding pattern outcomes following MSRF.

7. Unincorporated nucleotides are removed form the probe by passing the sample through a push column system.
8. Half the volume of the purified probe is denatured by heating at 95°C for 10 min and immediately cooled on ice. The radiolabeled probed is added to 20 mL of hybridization solution and incubated overnight with the nitrocellulose filter.
9. Following incubation the filter is stringently washed for 10 min in 10 mL of each 2X SSPE, 0.1% SDS, and 1X SSPE, 0.1% SDS.
10. The membrane is exposed to Kodak X-Omat film for 24–72 h at –70°C.

3.8. Interpretation of Results

Application of MSRF to the study of DNA methylation generates information regarding the methylation status of CpG islands within the DNA genome.

Using this approach, four different methylation states can be distinguished, as illustrated in **Fig. 1**. In outcome 1, PCR products are generated only in DNA from normal or tumor samples digested with *Mse*I alone (i.e., lanes 2 and 4). The absence of products in lanes 1 and 3 is due to *Bst*UI digestion at an unmethylated restriction site in the given genomic region.

In outcome 2, equally intense bands are observed in all four lanes; this can be caused by two situations. Firstly, the CGCG site(s) at the given genomic sequence is methylated in both tumor and normal DNA and, therefore will not be restricted by *Bst*UI. Alternatively, in this particular region of the genome there may be no *Bst*UI restriction sites and amplification of this region is permitted. Outcome 3 represents the case of an abnormal hypermethylation event. This situation is distinguished by the presence of a band, indicating PCR amplification, in the double-digested tumor DNA. If the hypermethylation at this region is exclusive to tumor DNA (outcome 3a), a corresponding band will not be observed following double digestion of normal DNA.

However, if this region of the DNA were partially methylated in normal DNA but becomes fully methylated in tumors, the banding pattern illustrated in outcome 3b would be generated. Finally, MSRF facilitates the detection of hypomethylated regions of the genome in tumors, as shown in outcome 4. In this situation a loss or reduction in band intensity in double-digested tumor DNA (lane 1) in comparison to the presence of a fully intense band in double-digested normal DNA (lane 3) is observed.

4. Notes

1. DNA extraction from cultured cells. To investigate the methylation status of CpG sites in cell-line models genomic DNA can be extracted from tissue-culture samples. Sufficient DNA may be obtained from 1×10^7 cells, which can be harvested by centrifugation at $800g$ for 10 min. The resulting cell pellet should be washed in 1X PBS and resuspended in 10 mL of cell-lysis buffer. DNA is extracted as described in **Subheading 3.1.**
2. Re-amplification of DNA extracted from bands excised from polyacrylamide gel can be problematic due to the low yield associated with 10-mer arbitrary primers. To overcome this problem several cycles of PCR may be necessary to obtain sufficient DNA for detection and cloning. Alternately, discrimination primers may be employed. These 20-mer primers consist of the original 10-mer sequence with additional 5′ sequences that have different overhang restriction sites. These primers improve amplification efficiency due to the extra stability associated with longer sequences. The design and application of these primers are described in detail by Zeng et al. *(13)*.

References

1. Bestor, T. H. (1990) DNA methylation: evolution of a bacterial immuno function into a regulator of gene expression and genome structure in higher eukaryotes. *Phil. Trans. R. Soc. Lond. B* **326,** 179–187.
2. Gruenbaum, Y., Stein, R., Cedar, H., and Razin, A. (1981) Methylation of CpG sequences in eukaryotic DNA. *FEBS Lett.* **124,** 67–71.
3. Bird, A. P. (1986) CpG rich islands and the functions of DNA methylation. *Nature* **321,** 209–213.
4. Riggs, A. D. and Pfeifer, G. P. (1992) X chromosome inactivation and cell memory. *Trends Genet.* **8,** 169–174.
5. Brandeis, M., Frank, D., Keshet, I., Seigfried, Z., Mendelsohn, M., Nemes, A., et al. (1994) Sp1 elements protect a CpG island from de novo methylation. *Nature* **371,** 435–438.

6. Stein, R., Sciaky-Gallili, N, Razin, A., and Cedar, H. (1982) Pattern of methylation of two genes coding for housekeeping gene functions. *Proc. Natl. Acad. Sci. USA* **79,** 2422–2426.
7. Issa, J.-P. J., Vertino, P. M., Wu, J., Sazawal, S., Celano, P., Nelkin, B. D., et al. (1993) Increased cytosine DNA-methyltransferase activity during colon cancer progression. *J. Natl. Canc. Inst.* **85,** 1235–1240.
8. Melki, J. R., Warnecke, P., Vincent, P. C., and Clark, S. J. (1998) Increased DNA methyltransferase expression in leukaemia. *Leukaemia* **12,** 311–316.
9. Baylin, S. B., Hoppener, J. W. M., deBustros, A., Steenberg, P. H., Lips, C. J. M., and Nelkin, B. D. (1986) DNA methylation patterns of the calcitonin gene in human lung cancers and lymphomas. *Cancer Res.* **46,** 2917–2922.
10. Herman, J. G., Jen, J., Merlo, A., and Baylin, S. B. (1996) Hypermethylation-associated inactivation indicates a tumour supressor role for p15^{INK4B}. *Cancer Res.* **56,** 722–727.
11. Huang, T. H. M., Laux, D. E., Hamlin, B. C., Tran, P., Tran, H., and Lubahn, D. B. (1997) Identification of DNA methylation markers for human breast carcinomas using the methylation-sensitive restriction fingerprint technique. *Cancer Res.* **57,** 1030–1034.
12. Southern, E. M. (1975) Detection of specific sequences among DNA fragments separated by gel electrophoresis. *J. Mol. Biol.* **98,** 503–506.
13. Zeng, M., Martsen, E. O., and Lapeyre, J. N. (1998) Re-amplification of short primer-generated bands from RAPD and methylation-sensitive restriction fingerprinting by discrimination primers. *Biotechniques* **24,** 402–403.

6

Restriction Landmark Genome Scanning

Joseph F. Costello, Christoph Plass, and Webster K. Cavenee

1. Introduction

Restriction landmark genomic scanning (RLGS) is a method that provides both a quantitative genetic and epigenetic (cytosine methylation) assessment of thousands of CpG islands in a single gel without prior knowledge of gene sequence *(1)*. The method is a two-dimensional separation of radiolabeled genomic DNA into nearly 2,000 discrete fragments that have a high probability of containing gene sequences and are ideal in length for cloning and sequence analysis. Genomic DNA is digested with an infrequently cutting restriction enzyme such as *Not*I, radiolabeled at the cleaved ends, digested with a second restriction enzyme, and then electrophoresed through a narrow, 60 cm-long agarose tube-shaped gel. The DNA in the tube gel is then digested by a third, more frequently cutting restriction enzyme and electrophoresed, in a direction perpendicular to the first separation, through a 5% nondenaturing polyacrylamide gel, and the gel is autoradiographed. Radiolabeled *Not*I sites are frequently used as "landmarks" because *Not*I can not cleave methylated sites and since an estimated 89% of *Not*I sites are within CpG islands *(2)*. Using a methylation-sensitive enzyme, the technique has been termed RLGS-M *(3)*. The resulting RLGS profile displays both the copy number and methylation status of the CpG islands. These profiles are highly reproducible and are therefore amenable to inter- and intra-individual DNA sample comparisons.

To increase the number of fragments analyzed by RLGS, the DNA samples can be processed with a different series of enzymes. The choice of a "landmark" enzyme is critical since this site determines the bias of the displayed fragments. To maintain a strong bias for CpG islands, landmark enzymes such as *Not*I, *Bss*HII, or *Eag*I are generally used. Alternatively, a different second and/or third restriction enzyme may be used along with same landmark enzyme to display a different subset of fragments.

Differences between RLGS profiles have been used to identify important genes involved in normal cellular processes and in disease states. Two novel imprinted genes, one encoding a ribonucleoprotein auxiliary factor and the second encoding Cdc25Mm, were isolated using this approach *(4,5)*. By determining the proportion of DNA fragments on RLGS profiles that display a potentially imprinted pattern, it has been possible to obtain an estimate of the total number of imprinted genes in the genome *(4)*. These genomic loci were identified as having a parent-of-origin-specific methylation pattern and indicated that a 50% change in the intensity of a single-copy DNA fragment was readily detectable in an RLGS profile. Such a reduction is also apparent in X-chromosome specific fragments derived from either males or females, since there is methylation-related inactivation of one X chromosome in the latter. Similarly, comparison of profiles from normal individuals to those from Down's Syndrome patients has revealed a proportional increase in the intensity of many chromosome 21 specific loci. Several chromosome 21 CpG islands were methylated on one copy of chromosome 21 and potentially represented an attenuation mechanism allowing for viability of a trisomy chromosome 21 fetus *(6)*. RLGS and standard positional cloning have also been combined to identify the gene responsible for cardiomyopathy in Syrian hamsters and to identify the mouse reeler gene *(7,8)*. Normal genetic variation among related individuals has also been detected by RLGS *(9)*. Thus, RLGS can be used for widespread methylation analysis of CpG islands and also to define a level of genetic or epigenetic change in cells.

This approach has been used to identify novel tumor-specific targets of DNA amplification, aberrant CpG island methylation, and repetitive sequences that are demethylated in human cancer and in experimentally induced rodent tumors *(10–12)*. For example, the gene encoding human cyclin-dependent kinase-6 (CDK6) was identified as a novel target of DNA amplification in human brain tumors *(12)*. Similarly, the tumor-suppressor gene, P16/INK4, as well as a variety of other CpG islands were identified as frequent targets of aberrant methylation in mouse liver tumors induced by tissue-specific expression of an SV-40 transgene *(10)*.

The chromosomal origin of the majority of DNA fragments displayed on RLGS profiles has been mapped by chromosome-assigned RLGS (CA-RLGS) *(13)*. CA-RLGS profiles were generated from flow-sorted human chromosomes and then each individual chromosome-specific profile was integrated into a total genomic DNA profile. In mouse, a more detailed linkage between 1045 DNA fragments displayed on RLGS profiles and specific loci within each chromosome has been generated from profiles derived from a panel of recombinant inbred mouse strains *(14)*.

RLGS-M has significant advantages when compared to polymerase chain reaction (PCR)-based global methylation-analysis techniques. First, there is a greater than 90% bias toward display of CpG island-containing DNA fragments when using *Not*I, which is critical since CpG islands are tightly linked with genes *(15)*. CpG island bias is especially important in examining cancer-cell genomes for altered methylation as the majority of such changes occur in noncoding, potentially nonfunctional regions of the genome. Certain repetitive elements in the genome are also rich in CpG dinucleotides and can be subjected to methylation changes in cancer cells *(11)*. These repetitive elements are displayed as high copy-number fragments on RLGS profiles, distinguishing them from single-copy CpG islands. Second, several thousand CpG island fragments can be analyzed simultaneously, whereas other methylation scanning methods visualize 10-fold fewer methylation sites. Third, quantitative information for each fragment can be derived directly from the profiles and compared to that obtained from many fragments that are of similar size and are invariant in intensity between samples. Fourth, since RLGS does not involve PCR amplification, the particular CpG islands that are displayed are not restricted by the limited ability of PCR enzymes to amplify GC-rich templates.

2. Materials
2.1. Isolation of Genomic DNA

1. Liquid nitrogen.
2. Mortar and pestle.
3. Heavy-duty aluminium foil.
4. Hammer.
5. 50-mL tubes.
6. Dialysis tubing (3/4 in × 25 ft) (Gibco BRL).
7. 100% ethanol.
8. Phenol.
9. Chloroform.
10. Isoamyl alcohol.
11. Dialysis clips/closures (Spectra/Por).
12. Proteinase K.
13. RNAse A (Boehringer Mannheim).
14. Sarkosyl (Fluka).
15. 8-hydroxyquinoline (Sigma).
16. Lysis buffer: 10 mM Tris-HCl, pH 8.0, 150 mM EDTA, pH 8.0, and 1% sarkosyl.
17. PCI: phenol : chloroform : isoamylalcohol in the ratio 50 : 49 : 1.

2.2. Enzymatic Processing of Genomic DNA

1. Wide-bore pipet tips.
2. Dithiothreitol (DTT).
3. Triton X-100.
4. Bovine serum albumin (BSA).
5. dGTPαS, dCTPαS, ddATP, ddTTP, ddGTP, ddCTP (Pharmacia).
6. DNA polymerase I (3.5 U/μL, Boehringer Mannheim).
7. Sequenase ver 2.0 (13 U/μL, USB/Amersham).
8. Not I (10 U/μL).
9. EcoR V (10 U/μL).
10. Hinf I (70 U/μL) (Promega).
11. [α-^{32}P]-dGTP: 20 mCi/mL, 6000 Ci/mmol (New England Nuclear).
12. [α-^{32}P]-dCTP: 10 mCi/mL, 6000 Ci/mmol (Amersham).
13. 10X Buffer 1: 500 mM Tris-HCl, pH 7.4, 100 mM MgCl$_2$, 1 M NaCl, 10 mM DTT (store at –20°C).
14. 20X Buffer 2: 3 M NaCl, 0.2% Triton X-100, 0.2% BSA (store at –20°C).
15. Blocking buffer: 1 μL 10X buffer 1, 0.1 μL 1 M DTT, 0.4 μL each of 10 μM dGTPαS, 10 μM ddATP, 10 μM ddTTP, and 0.2 μL 10 mM dCTPαS (make stock and store in aliquots at –20°C)
16. Second enzyme digestion buffer: 1 μL 1 μM ddGTP, 1 μL 1 μM ddCTP, 4.4 μL ddH$_2$O, 1.2 μL 100 mM MgCl$_2$.
17. 6X Loading dye, first-dimension; 0.25% bromophenol blue (BPB), 0.25% xylene cyanol (XC), 15% Ficoll type 400.

2.3. First-Dimension Gel Set-Up and Electrophoresis

1. First-dimension gel apparatus (C.B.S Scientific and EverSeiko Corp., Tokyo).
2. PFA-grade Teflon tubing for the first-dimension gel: 2.4 mm i.d., 3.0 mm o.d., 10 M, sufficient for at least 16 gels; PFA 11 thin-wall, natural (American Plastic, Columbus, OH).
3. Glass tubes (4 mm i.d., 5 mm o.d., and 60 cm) with a tapering at the top end extending over 1.2 cm to a final dimension of 3 mm i.d.).
4. Two-way stopcocks (4–8).
5. Flexible Tygon tubing (3/16 in. i.d., 1/4 in. o.d., VWR).
6. Seakem GTG agarose (do not substitute).
7. 20X Boyer's buffer: 1 M Tris, 360 mM NaCl, 400 mM sodium acetate, 40 mM EDTA, autoclave. Important: Boyer's buffer is used at 2X concentration for the gel and running buffer.
8. Salmon sperm DNA (500 μg/mL).
9. 20% Trichloroacetic acid (TCA).
10. Whatman GF/F filter.
11. First-dimension gel (0.8%); 0.48 gm agarose, 60 mL 2X Boyer's buffer.

2.4. In-Gel Digest

1. Digest tubing: PFA-grade teflon, 9, thin-wall, natural, 2.7 mm i.d. and approx 3.3 mm o.d. (American Plastic).
2. 10-mL syringe.
3. 10X Buffer K; 200 mM Tris-HCl, pH 7.4, 100 mM MgCl$_2$, 1 M NaCl, autoclave.
4. *Hinf* I (70 U/µL; Promega).

2.5. Second-Dimension Electrophoresis

1. Five glass plates.
2. 8 spacers.
3. 1 Plexiglas sheet (C.B.S. Scientific).
4. Flexible tubing (1/8 in. i.d., 3/16 in. o.d., VWR).
5. Three-way stopcock.
6. 10-mL syringes.
7. 60-mL syringes.
8. "Plastic" tape (Scotch brand, do not substitute).
9. Whatman paper (3 for each gel, 13.5 in × 16.5 in).
10. 10X TBE, pH 8.3.
11. Connecting agarose; 1X TBE, pH 8.3, 0.8% Seakem GTG agarose.
12. Second-dimension loading dye; TE, pH 8.0, 0.25% BPB, 0.25% XC.
13. 5% Nondenaturing polyacrylamide gel; 1X TBE, pH 8.3, 96.9 gm acrylamide, 3.3 gm bis-acrylamide, 1.3 gm ammonium persulfate (APS) in a total volume of 2 L. Add 700 mL TEMED before pouring the gel. Makes 4 gels (one second-dimension apparatus).

2.6. Analysis of RLGS Profiles

1. Light box.
2. Clear acetate sheets.

2.7. Cloning of DNA Fragments

1. *Not*I, *Eco*RV and *Hinf* I (Promega).
2. Sequenase ver. 2.0 (USB).
3. T4 DNA ligase, (New England Biolabs).
4. Taq DNA polymerase (Boehringer Mannheim).
5. TE buffer, pH 8.0.
6. 0.1% Triton X-100.
7. PCI.
8. 3 M sodium acetate.
9. 7.5 M ammonium acetate.
10. Ethanol (100%).

11. 10 m*M* dNTP mix: [α-^{32}P]-dGTP: 20 mCi/mL, 6000 Ci/mmol (New England Nuclear); [α-^{32}P]-dCTP: 10 mCi/mL, 6000 Ci/mmol (Amersham), 50% PEG 6000
12. *Not*I-restriction trapper (Japan Synthetic Rubber Co., Tokyo).
13. Electroeluter (Biocraft, Japan).
14. Primers (not phosphorylated).
    F-1: 5′-TACGTCGAACGCCAGGGTTTTC-3′
    F-2: 5′-CCAGTCACGACGCGGCCGC-3′
    R-1: 5′-AGCAGTCCGGATATCCTGGTG-3′
    R-2: 5′-GATATCCTGGTGCAGTACAGANTC-3′
    NOTI-1: 5′-TACGTCGAACGCCAGGGTTTTCCCAGTCACGACGC-3′
    NOTI-2: 5′-GGCCGCGTCGTGACTGGGAAAACCCTGGCGTTCGACGTA-3′
    HINFI-1: 5′-ANTCTGTACTGCACCAGGATATCCGGACTGCT-3′
    HINFI-2: 5′-AGCAGTCCGGATATCCTGGTGCAGTACAG-3′
15. 1% Linearized acrylamide; Add 1 g acrylamide, 0.01 g APS, and 15 mL TEMED in 10 mL ddH$_2$O. Dilute to appropriate concentration by adding 90 mL ddH$_2$O after acrylamide is solidified.
16. 6X loading dye; 0.25% bromophenol blue, 0.25% XC, 15% Ficoll type 400 (Pharmacia) in water.
17. Elution dye: mix 95 µL 7.5 *M* ammonium acetate and 5 µL 6X loading dye

3. Methods

3.1. Isolation of Genomic DNA

DNA quality is a critical parameter for generating high-quality RLGS profiles. Small amounts of degraded DNA can cause a diffuse background. Therefore, tissue and cell pellets should be snap-frozen in liquid nitrogen and stored at –80°C prior to isolation. The DNA isolation procedure follows published protocols *(16)* with several modifications.

1. Add 2 mL lysis buffer (without proteinase K) to 100–300 mg tissue in a 50 mL Falcon tube and freeze in liquid nitrogen. Expel the frozen tissue and buffer, wrap it in aluminium foil, and quickly break into pieces with a hammer. Keep the foil/tissue cold by submerging it in liquid nitrogen.
2. Transfer the tissue to a mortar (pre-chilled in liquid nitrogen) and grind to a powder with a pre-chilled pestle. Transfer the tissue powder into a 50-mL tube and store in liquid nitrogen until all samples are processed.
3. Add 15–25 mL of lysis buffer and proteinase K (0.1 mg/mL final conc. in lysis buffer). Mix gently using a glass rod or 1-mL disposable pipet. Incubate at 55°C for 20 min, mixing very gently every 5 min.
4. Cool the lysed samples on ice for 10 min. Add an equal volume of PCI. Rotate tubes gently for 30–60 min.

5. Centrifuge for 30 min at 2500 rpm and transfer the DNA (aqueous phase) to a 50-mL tube using a wide-bore pipet. Repeat the PCI extraction.
6. Transfer the DNA into dialysis tubing and dialyze against 4 L of 10 mM Tris-HCl, pH 8.0, for 2 h. Transfer the tubing into fresh 10 mM Tris-HCl and dialyze overnight at room temperature. Dialyze in fresh 10 mM Tris for an additional 2 h.
12. Transfer the DNA to 50-mL tubes and add RNase A to a final concentration of 1 mg/mL. Incubate at 37°C for 2 h.
13. Add 2.5 vol 100% ethanol to the DNA and rotate gently.
14. Transfer the pellet to a microfuge tube, centrifuge briefly, and remove the remaining ethanol. Do not over-dry the pellet. For smaller amounts of tissue, it may be necessary to collect the DNA by centrifugation at 3000 rpm. Discard the ethanol and briefly air-dry the pellet.
15. Resuspend the DNA to a final concentration of approx 1 mg/mL (may take several days). To check for DNA degradation, electrophorese a small amount of the DNA on a standard 0.8% agarose gel. DNA isolated in this manner has an average size of 200–300 kb and should be transferred with positive displacement pipets.

3.2. Enzymatic Processing of Genomic DNA

Genomic DNA is first blocked at sheared sites by the addition of dideoxynucleotides and sulfur-substituted nucleotides. The DNA is digested with an infrequent cutting restriction enzyme, end-labeled at the restriction sites, and then digested with a second restriction enzyme. To prevent nonspecific shearing of the DNA during this procedure, wide-bore pipet tips should be used for transferring the DNA, and all reactions must be mixed by stirring, rather than pipeting. The protocol below describes in detail the procedure and solutions for the most frequently used restriction enzyme set (*Not*I, *Eco*RV, *Hinf*I), although many different methylation-sensitive and insensitive restriction enzymes may be used (*see* **Table 1**).

1. In a 1.5-mL tube add 7 µL genomic DNA (0.2–0.6 µg/µL), 2.5 µL blocking buffer, and 0.5 µL DNA polymerase I, mix thoroughly by stirring, and incubate reaction at 37°C for 20 min. (Transfer the DNA with a wide-bore pipet tip. Do not pipet to mix. Using a master mix increases the uniformity among samples.)
2. Incubate the reaction at 65°C for 30 min to inactivate the polymerase. Cool the reaction on ice for 2 min, centrifuge briefly at high speed.
3. Add to the sample 8 µL 2.5X buffer 2, 2 µL *Not*I (10 U/µL) and stirred to mix thoroughly. Incubate at 37°C for 2 h.
4. Add to the sample 0.3 µL 1 M DTT, 1 µL [α-^{32}P]-dGTP, 1 µL [α-^{32}P]-dCTP, and 0.1 µL Sequenase ver 2.0 (13 U/µL; use a master mix). Mix thoroughly by stirring and incubate at 37°C for 30 min.

**Table 1
Examples of Additional Enzyme Combinations**

Enzyme Combination	Reference
NotI/EcoRV/HinfI	(1)
NotI/PvuII/PstI	(10)
NotI/PstI/PvuII	(8)
NotI/BamHI/HinfI	(8)
BssHII/PvuII/PstI	(8)
BssHII/BamHI/EcoRI	(8)
BssHII/BamHI/EcoRV	(8)
PacI/EcoR I/MboI (methylation-insensitive)	(21)

5. Add to the sample 7.6 µL second-digestion buffer and 2 µL *Eco*RV (10 U/µL). Mix by stirring and incubate at 37°C for 1 h. Cool on ice and add 7 µL of 6X first-dimension loading dye. To confirm that the DNA was digested to completion, check 2.5 µL of the reaction on a standard 0.8 % agarose gel.

3.3. First-Dimension Gel Set-Up and Electrophoresis

3.3.1. Measurement of DNA Amount

A critical factor in the first-dimension electrophoresis is loading the correct amount of labeled DNA on the gel. Loading more than 1.5 µg may cause smearing of high molecular-weight DNA fragments. Loading less than 1 µg may result in a less intense profile, but longer film exposure times may compensate for this.

For visual determination of the concentration of the DNA in each sample, co-electrophorese control restriction enzyme-digested DNA in the range of 0.2–1 µg on a standard 0.8% agarose gel. Approximate the microliter amount of the sample that will contain 1.5 µg. This is the maximum amount that should be loaded on the first-dimension gel. Alternatively, the amount of sample to be loaded on the first-dimension gel can be determined by loading a constant amount of incorporated radiolabel (**steps 1–3** below).

1. Mix 2 µL of each sample with 100 µL salmon sperm DNA (500 µg/mL) and add 100 mL 20% TCA. Incubate for 10 min on ice.
2. Filter each sample through a Whatman GF/F filter. Wash each filter twice with 10 mL 20% TCA.
3. Measure the dpms on each filter using a scintillation counter.

Two-Dimensional DNA Profiling

For mouse genomic DNA, loading 130,000 dpm is sufficient. For human genomic DNA, 450,000 dpm should be loaded. These values may vary with the quality of the DNA.

3.3.2. First-Dimension Gel

1. Gel holder: Using a sharp razor, cut one end of the Teflon tubing at an angle to make a bevel.
2. Feed the beveled end into the glass rod until it protrudes slightly from the tapered end. Using a hemostat, pull the beveled end up through the tapered end of the glass rod until it protrudes 2–4 cm.
3. Cut the tubing horizontally at the same end, leaving a 2-mm protrusion (this is the top of the gel holder).
4. Cut the opposite end horizontally to leave a 5–6 cm protrusion from the glass rod.
5. Invert the gel holder and press the top protruding end firmly against a hot metal surface (metal spatula heated by a Bunsen burner) to fold the edges of the Teflon outward onto the rim of the glass support, making sure to avoid folding the edges inward and sealing the tubing.
6. Pull a rubber stopper with cored center over the top end of the gel holder until it is just past the taper of the glass rod. It is essential that all tubing is clean and free of liquid and particulate matter. The Teflon tubing should be rinsed by suctioning through ddH_2O and then dried by continued application of the vacuum.
7. Attach a two-way stopcock to a 10 mL syringe and then to the gel holder via 2–3 cm of flexible tubing. Adjust the stopcock valve to the open position.
8. In a clean 200-mL glass bottle, add 60 mL 2X Boyer's buffer and 0.48 gm Seakem GTG agarose (0.8 %). Weigh the solution. Microwave until the agarose is dissolved, stopping occasionally to swirl the contents and to avoid boil-over. Weigh the solution and add ddH_2O to return to the starting weight. Equilibrate the gel solution to 55°C in a water bath.
9. With the stopcock valve in the open position, lower the protruding Teflon tube into the molten agarose solution. Suction the gel solution into the gel holder until the gel solution has reached 1–2 cm from the top of the gel holder and then close the stopcock valve. Keeping the gel upright, suspend the gel from a ring stand. Add a drop of gel solution to the bottom of each Teflon tube to allow for the slight gel shrinkage during the drying period. Allow the gel to solidify for a minimum of 20 min (60 mL gel solution is sufficient for 8 gels).
10. Open the stopcock valve and remove the syringe and connecting tubes from each gel.
11. After adding 2X Boyer's buffer to the bottom of the first-dimension apparatus (to approx 5 cm from the bottom), lower the gels into the first-dimension gel apparatus, seating the rubber stopper firmly into the appropriate holes in the top portion of the apparatus.
12. Fill the top chamber with 2X Boyer's buffer until the tops of the gels are submerged. Remove air bubbles from the space between the top of the gel holder

and the top of the gel. The sample will not run properly if the gel or the loading well contains bubbles or particulate matter.

13. Load an appropriate amount of sample onto each gel. Electrophorese at 110 V for 2 h, and then 230 V for approx 24 h (or until the BPB dye has reached 10 cm from the bottom of the lower buffer chamber).

3.4. In-Gel Digest

1. Remove buffer and gel holders from the first-dimension apparatus. Extrude the gel into a pan containing 1X buffer K by forcing the gel out through the bottom of the gel holder. This is accomplished using a 1-mL syringe fitted with a pipet tip and filled with buffer K. Firmly insert the tip into the top of the gel holder and depress the plunger until the gel begins to come out through the bottom of the gel holder. Carefully replace the 1-mL syringe with a 5-mL syringe; depress the plunger until the entire gel is expelled.
2. With a razor, cut a bevel in the low molecular-weight end of the gel and cut horizontally at the high molecular-weight end so that the gel is approx 43 cm in length. The gel length is now the same as the width of the second-dimension gel.
3. Place each gel into a separate 50-mL tube containing 40 mL of 1X buffer K. Incubate for 10 min at room temperature. Carefully pour off the buffer and incubate in 1X buffer K for an additional 10 min. (The gel may be transferred by carefully looping it onto gloved fingers.)
4. Carefully pour the buffer K and gel into a pan containing fresh buffer K. Using a 10-mL syringe attached to restriction-digest tubing (via a 1–2 cm segment of flexible tubing), suction the gel into the digest tubing, low molecular-weight (beveled) end first.
5. The gel is suctioned into the digest tubing by placing the end of tubing in line with the beveled end of the gel and pulling the syringe plunger. Be careful to stop once the gel has completely entered the tubing. Carefully position the tubing vertically, with the syringe at the bottom. Suction any remaining buffer from the tubing into the syringe. Do not continue suctioning if the gel blocks the syringe opening. If this occurs, depress plunger gently and force gel away from syringe opening. Detach the syringe, expel buffer, and reattach.
6. In a clean tube, make a 1.6 mL mix of 1X restriction enzyme buffer, 0.1% BSA, and 750 U of *Hinf* I restriction enzyme. Place the open end of the digest tubing into the tube containing restriction-digestion solution, now holding the syringe end up, apply suction until a small amount of digestion solution appears in the syringe. Carefully remove the digest tubing and orient both ends upward in a U-shape. Remove the syringe and attach the two ends of the tubing to form a closed circle. Place in a moist chamber and incubate at 37°C for 2 h. Making a master mix of the digest solution works well. Avoid bubbles in the digest tubing as these may interfere with complete digestion. In general, the restriction enzyme is in excess and incomplete digests are very rare.

3.5. Second-Dimension Electrophoresis

1. Assembly of the second-dimension gel apparatus: all glass plates should be cleaned thoroughly and wiped with 95% ethanol. The nonbeveled face of each plate should be coated with Gelslick or Sigmacote (only once every 10 uses).
2. Lay the back half of the apparatus horizontally on a table top with the upper buffer chamber hanging over the table edge.
3. Insert the two small clear plastic blocks at the bottom corners of each apparatus. Place a glass plate in the apparatus, beveled edge facing upward and near the upper buffer chamber, followed by two spacers, one along each side. Add glass plates and spacers in this manner until the fifth plate has been added.
4. After the third plate, slide flexible Tygon tubing down the side channel of the apparatus, with a bevel cut in the leading end of the tubing. Cut the other end, leaving approx 10 cm protruding from the apparatus.
5. Place the Plexiglas "filler" sheet over the fifth glass plate. Position the front half of the apparatus by aligning the screw holes of the front and back half. Secure with the Teflon screws. Seal the oblong oval "windows" at the lower, front face with Plastic tape. Stand the apparatus upright in the lower buffer chamber.
6. Using a three-way stopcock, attach the gel apparatus tubing in series with a 2-L reservoir and attach a 60-ml syringe to the remaining stopcock outlet. The tubing should be attached to the 2-L reservoir through a bottom drain (A 2-L graduated cylinder works well).
7. Secure the reservoir above the gel apparatus to allow gravity flow. Adjust the stopcock valve to allow liquid to flow between the 2-L reservoir and the 60-mL syringe. Once the TEMED has been added, pour the acrylamide solution into the 2-L reservoir. Pull the syringe plunger down to the 50 mL mark. Depress the plunger to push the air out of the upper tubing. Once all air has been removed, adjust the valve so that all three ports are open. Acrylamide will flow into the apparatus, filling all four gels simultaneously from the bottom upward. Stop the flow when the level reaches 3 mm from the top edge of the glass plates.
8. Allow the solution to settle for 2–3 min. If the acrylamide level drops, continue flow briefly. Immediately add 1 mL of isopropyl alcohol along the top edge of each gel. After the valve leading to the gel apparatus has been closed, detach the syringe and reservoir. Once the acrylamide has polymerized, the gels may be stored overnight by adding 1X TBE to the upper reservoir. Immediately before use, rinse wells thoroughly with water and then dry.
9. Gently separate the ends of the digest tubing and extrude the first-dimension gel into a pan containing 1X TBE, pH 8.3. The gel may slide out by gravity or may require gentle liquid pressure.
10. Transfer the gel to a 50-mL tube containing 40 mL 1X TBE, pH 8.3. Incubate for 10 min at room temperature, replace with fresh TBE, and incubate for an additional 10 min.

11. Place each first-dimension gel in a horizontal position across the beveled edge of each glass plate.
12. Once all gels are in place, fill the space between the agarose gel and the top of each polyacrylamide gel with molten 0.8% agarose (equilibrated to 55°C).
13. Use an 18-gauge needle attached to a 10-mL syringe to add the connecting agarose. Be sure to avoid having bubbles between the first and second dimension gels. Allow connecting agarose to solidify for 10–15 min, add 250 mL second-dimension loading dye along the length of each gel.
14. Add 1X TBE, pH 8.3, to the upper and lower buffer chambers and electrophorese at 100 V for 2 h and then at 150 V for approx 24 h (or until the BPB reaches the bottom of the gel).
15. Remove buffers and disassemble apparatus.
16. Lift each gel from the plates by overlaying with Whatman paper cut to size for autoradiographic or phosphorimager cassettes. Trace the perimeter of the paper with edge of a plastic ruler, removing excess gel.
17. Carefully peel back and lift Whatman paper and gel. Place gel side up on second Whatman paper. Overlay with Saran wrap, add third Whatman paper to top and fold edges of Saran wrap over top Whatman. In the same orientation, place in a gel drier for 1 h at 80°C while applying a vacuum.
18. Remove lower and upper Whatman paper, fold Saran wrap under remaining paper and expose to X-ray film (BioMax MS).

3.6. Analysis of RLGS Profiles

Using direct visual assessment of profiles has proven very reliable. There is a 100% concordance between alterations of DNA fragments detected in RLGS profiles and, following cloning, alteration demonstrated by Southern blotting. Comparisons by computer or by visual assessment are facilitated by the fact that RLGS profiles from different tissues or from different individuals are identical at the majority of loci. Visual assessment is performed by overlaying two RLGS profiles on a light box and comparing relative intensities of fragments. Overlaying "master" profiles (profiles used as a standard for comparison) with clear acetate sheets allows one to mark differences or similarities between multiple profiles and to generate cumulative data sets. This is also a convenient form in which to retain a usable record of the analysis results, and is a standard that has allowed the sharing of results between labs. Several computer-assisted analysis systems, which were originally designed for analysis of two-dimensional (2-D) protein gels, have been developed for the 2-D DNA analysis *(17,18)*.

One of the disadvantages of using RLGS for global methylation analysis of CpG islands, at least in the analysis of tumor tissue, is that a loss of a fragment from an RLGS profile could be due to either deletion or methylation. Once the fragment is cloned, these possibilities are distinguished by Southern blotting.

It has been our experience that the vast majority of "loss" events detected by RLGS in human tumors are due to methylation. To tailor the profiles for detection of genetic loss events only, methylation-insensitive landmark enzymes are used (*see* **Table 1**).

The identification and cloning of DNA fragments that display a tumor-specific increase in intensity has led to two very different findings, which are indistinguishable until the DNA fragments are cloned and tested as probes on Southern blots. The increased intensity in some cases corresponded to gene amplification *(12)* while for others identified DNA fragments from repetitive sequences that had become demethylated in a tumor-specific manner *(11)*. Thus, the interpretation of tumor-specific intense fragments on profiles must be refined by cloning and Southern-blot analysis. As additional fragments are sequenced, the profiles will become even more informative.

3.7. Cloning of DNA Fragments

The following cloning protocol combines the advantages of the two published cloning procedures: the restriction trapper-based direct cloning of *Not*I/*Hinf*I *(19)* and the PCR-mediated method *(20)*. Standard RLGS gels contain a large amount of unlabeled background *Hinf*I or *Hinf*I/*Eco*RV fragments, which could compete with target *Not*I/*Hinf*I fragments in the cloning process. A protocol was developed that eliminates more than 90% of the *Eco*RV/*Eco*RV fragments, while retaining *Not*I/*Eco*RV fragments. RLGS profiles generated with purified *Not*I/*Eco*RV fragments are indistinguishable from total genomic DNA profiles. Using the human genome sequence, it is also possible to identify the sequence and chromosomal position of fragments through in silico restriction digests.

3.7.1. Purification of NotI/EcoRV Fragments Using the NotI Restriction Trapper

1. Digest 100 µg genomic DNA with *Not*I and *Eco*RV in a total volume of 200 µL. Incubate the reaction for 2 h at 37°C.
2. Extract the DNA with an equal volume of PCI. Add 2.5 volumes of 100% ethanol, mix, and centrifuge for 5 min. Remove the ethanol and wash the pellet with 70% ethanol. Resuspend the DNA in 50 µL TE, pH 8.0. Adjust the concentration to 2 µg/µL.
3. In a separate tube, add 260 µL TE to 140 µL 10% restriction trapper solution. Centrifuge 5 min at full speed, discard the supernatant, and dissolve the pellet in 280 µL TE.
4. To ligate the *Not*I fragments to the restriction trapper, mix 50 µL (100 µg) of genomic DNA and 20 µL 5% *Not*I restriction trapper. Bring the reaction volume to 120 µL with 1X ligase buffer and 1,400 U T4-DNA ligase. Mix well and

add 10% PEG-6000 (final conc.). After 30 min mix the reaction by pipetting. Incubate overnight at 18°C.
5. Bring the reaction to 600 µL with 1X restriction buffer and add 100 U of EcoRV. Incubate the reaction at 37°C for 1 h.
6. Add 60 µL 0.1% Triton X-100 to the digestion mix, centrifuge 5 min, discard the supernatant and resuspend the pellet in 50 µL 1X EcoRV buffer. Add 4 µL EcoRV (10 U/µL) and incubate at 37°C for 1 h.
7. Add 200 µL ddH$_2$O and 40 µL 0.1% Triton X-100 and mix. Centrifuge for 5 min, discard supernatant, and dissolve the pellet in 200 µL 1X NotI restriction buffer. Mix carefully, centrifuge, and remove all supernatant (it may contain trace amounts of EcoRV fragments).
8. Dissolve the pellet in 100 µL 1X NotI restriction buffer, add 50 U NotI and incubate for 2 h at 37°C. Centrifuge for 5 min, (save the supernatant) and wash the pellet with 200 µL TE buffer containing 0.05 % Triton X-100. Centrifuge for 5 min and combine supernatants.
9. Add an equal volume of PCI, mix, and centrifuge for 5 min. Transfer the supernatant to a fresh tube and extract with an equal volume of chloroform: isoamyl alcohol. To precipitate the DNA, add 20 µL 3 M sodium-acetate, 1 µL linearized acrylamide, and 550 µL ethanol (100%). Mix well, incubate at –80°C for 30 min and then centrifuge for 30 min. Rinse the pellet with 70% ethanol, dry, and resuspend in 13 µL TE. Measure the DNA concentration (approx 100–600 ng total).

3.7.2. Two-Dimensional Separation of Purified Fragments

Since the 5′ overhangs created by NotI digestion are "filled in" during the labeling step, standard DNA "analysis" gels are not suitable for cloning with NotI adapters. To generate profiles suitable for such cloning, only 1/5 of the restriction trapper-purified DNA is subjected to labeling and is then mixed with the remaining 4/5 unlabeled DNA prior to loading the first-dimension agarose gel.

1. For labeling, 3 µL of the purified NotI/EcoRV fragments are added to a 10 µL reaction containing 1X sequenase buffer, 20 µCi [α-^{32}P]-dGTP, 10 µCi[α-^{32}P]-dCTP, and 13 U Sequenase ver. 2.0. Incubate the samples at 37°C for 30 min. Inactivate the enzyme at 65°C for 30 min.
2. Add 9 µL of the unlabeled NotI/EcoRV sample and 5 µL of 6X loading dye.
3. Add the sample to the first-dimension gel and proceed as described in **Subheading 4**.
4. Dry the gels at 65°C and expose to X-ray film overnight. The film should be stapled to the gel during exposure to allow for precise orientation and reattachment of the developed film to the gel. DNA fragments of interest are excised from the gel using a scalpel and collected in a 1.5-mL tube.

3.7.3. Elution of the DNA

Use method 1 (**Subheading 3.7.3.1.**) or method 2 (**Subheading 3.7.3.2.**).

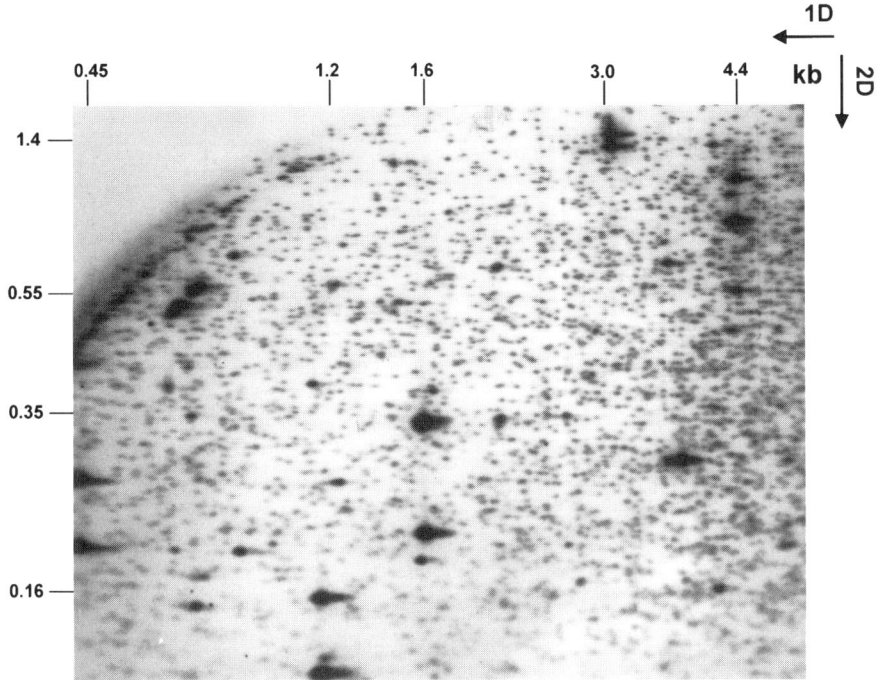

Fig. 1. RLGS profile of normal human cerebellum DNA using the enzyme combination *Not*I/*Eco*RV/*Hinf*.

3.7.3.1. METHOD 1

1. Remove the piece of X-ray film and Saran wrap from the excised gel slice and soak the gel in 20–40 µL TE buffer. Place the gel slice into the precast molding of the electroeluter.
2. Add 1X TAE buffer to the buffer chambers and add 100 µL elution dye into the V-shaped slots. Apply current to trap the DNA in the V-shaped slot.
3. Recover the elution dye (approx 300 µL) and add extract with an equal volume of PCI. Precipitate the DNA by adding 2 µL linearized acrylamide and 2.5-vol 100% ethanol (100%) and incubate at –70°C. Centrifuge and remove the supernatant. Dry the pellet and dissolve in 10 µL TE-buffer.

3.7.3.2. METHOD 2

1. Remove the piece of X-ray film and Saran wrap from the excised gel slice. Wet the gel slice with 20 µL of elution buffer (0.5 M ammonium acetate, 1 mM EDTA). Gently lift the rehydrated gel off of the paper.
2. Cut the gel slice into smaller pieces with a scalpel. Add 150 µL elution buffer to the wetted gel slice. Place in a shaking incubator at 37°C overnight.

3. Centrifuge and transfer supernatant to a clean tube. Rinse acrylamide with 50 µL elution buffer and combine with supernatant.
4. Add 1 µL glycogen (20 µg/µL) and 2 vol 100% ethanol; chill for several hours or overnight at −70°C.
5. Centrifuge at 4°C for 30 min, remove ethanol, and dry pellet. Resuspend in 10 µL TE.

3.7.4. Adaptor Ligation and PCR Amplification

If sufficient DNA can be eluted from the restriction trapper gels direct cloning into a vector prepared with *Not*I/*Hinf*I ends is possible. For cloning from limited amounts of DNA, the following protocol should be used.

1. *Not*I and *Hinf*I adapters are prepared by annealing primer sets *Not*I-1 with *Not*I-2 and *Hinf*I-1 with *Hinf*I-2, respectively. Equimolar amounts of each primer are mixed in a total volume of 100 µL and heated to 95°C for 5 min. The reaction is cooled slowly over 3 h to 25°C.
2. Set up a 10 µL ligation reaction containing 2.5 µL of the eluted DNA fragments, 1 mM *Not*I and 1 mM *Hinf*I adapters, 350 U T4-DNA ligase, and 1X ligase buffer. Incubate overnight at 14°C.
3. Dilute the ligation reaction by adding ddH$_2$O to a total vol of 40 µL.
4. Two rounds of PCR amplification are performed using nested primer sets. The first PCR reaction is carried out using primers F-1 and R-1 in a standard PCR reaction, except containing 3 mM MgCl$_2$. After an initial denaturation step, 10 cycles of 90°C for 1 min, 65°C for 1 min, and 72°C for 1 min are performed, followed by a 10 min final extension at 72°C.
5. One-fifth of the first PCR reaction is used as a template for the second PCR with primers F-2 and R-2. This reaction is performed in 3 mM MgCl$_2$ for 30 cycles using the same cycling conditions.
6. The PCR products are gel-purified and cloned into an appropriate vector.
7. Since both *Not*I/*Hinf*I fragments and *Hinf*I/*Hinf*II fragments may be coamplified, we recommend screening bacterial colonies by hybridization with radiolabeled F-2.

Acknowledgments

We would like to thank Dr. Gavin P. Robertson for critical review of this chapter, Dr. Michael Fruehwald for the RLGS profile, and Dr. Yoshihide Hayashizaki for making possible the transfer of this technology to our laboratory.

References

1. Hatada, I., Hayashizaki, Y., Hirotsune, S., Komatsubara, H., and Mukai, T. (1991) A genomic scanning method for higher organisms using restriction sites as landmarks. *Proc. Natl. Acad. Sci. USA* **88,** 9523–9527.
2. Lindsay, S. and Bird, A. P. (1987) Use of restriction enzymes to detect potential gene sequences in mammalian DNA. *Nature* **327,** 336–338.

3. Kawai, J., Hirose, K., Fushiki, S., Hirotsune, S., Ozawa, N., Hara, A., et al. (1994) Comparison of DNA methylation patterns among mouse cell lines by restriction landmark genomic scanning. *Mol. Cell. Biol.* **14**, 7421–7427.
4. Hayashizaki, Y., Shibata, H., Hirotsune, S., Sugino, H., Okazaki, Y., Sasaki, N., et al. (1994) Identification of an imprinted U2af binding protein related sequence on mouse chromosome-11 using the RLGS method. *Nature Genet.* **6**, 33–40.
5. Plass, C., Shibata, H., Kalcheva, I., Mullins, L., Kotelevtseva, N., Mullins, J., et al. (1996) Identification of grf1 on mouse chromosome 9 as an imprinted gene by RLGS-M. *Nature Genet.* **14**, 106–109.
6. Kuromitsu, J., Yamashita, H., Kataoka, H., Takahara, T., Muramatsu, M. Sekine, T. et al. (1997) A unique downregulation of h2-calponin gene expression in Down's syndrome: a possible attenuaton mechanism for survival by methylation at the CpG island in the trisomic chromosome 21. *Mol. Cell. Biol.* **17**, 707–712.
7. Okazaki, Y., Okuizumi, H., Ohsumi, T., Nomura, O., Takada, S., Kamiya, M., et al. (1996) A genetic linkage map of the syrian hamster and localization of the cardiomyopathy locus on chromosome 9qa2.1-b1 using RLGS spot-mapping. *Nature Genet.* **13**, 87–90.
8. Okazaki, Y., Hirose, K., Hirotsune, S., Okuizumi, H., Sasaki, N., Ohsumi, T., et al. (1995) Direct detection and isolation of restriction landmark genomic scanning (RLGS) spot DNA markers tightly linked to a specific trait by using the spot-bombing method, *Proc. Natl. Acad. Sci. USA* **92**, 5610–5614.
9. Kuick, R. M., Asakawa, J., Neel, J. V., Kodaira, M., Satoh, C., Thoraval, D., et al. (1996) Studies of the inheritance of human ribosomal DNA variants detected in two-dimensional separations of genomic restriction fragments. *Genetics* **144**, 307–316.
10. Akama, T. O., Okazaki, Y., Ito, M., Okuizumi, H., Konno, H., Muramatsu, M., et al. (1997) Restriction landmark genomic scanning (RLGS-M)-based genome-wide scanning of mouse liver tumors for alteratons in DNA methylation status. *Cancer Res.* **57**, 3294–3299.
10a. Costello, J. F., Früwald, M. C., Smiraglia, D. J., Rush, L. J., Robertson, G. P., Gao, X., et al. (2000) Aberrant CpG island methylation has non-random and tumour type-specific patterns. *Nature Genetics* **25**, 132–138.
11. Miwa, W., Yashima, K., Sekine, T., and Sekiya, T. (1995) Demethylation of a repetitive DNA sequence in human cancers. *Electrophoresis* **16**, 227–232.
12. Costello, J. F., Plass, C., Arap, W., Chapman, V. M., Held, W. A., Berger, M. S., et al. (1997) Cyclin dependent kinase 6 (CDK6) amplification in human gliomas identified using two dimensional separation of genomic DNA. *Cancer Res.* **57**, 1250–1254.
13. Yoshikawa, H., Delamonte, S., Nagai, H., Wands, J. R., Matsubara, K., and Fujiyama, A. (1996) Chromosomal assignment of human genomic Not I restriction fragments in a two-dimensional electrophoresis profile. *Genomics* **31**, 28–35.
14. Hayashizaki, Y., Hirotsune, S., Okazaki, Y., Shibata, H., Akasako, A., Muramatsu, M., et al. (1994) A genetic linkage map of the mouse using restriction landmark genomic scanning (RLGS). *Genetics* **138**, 1207–1238.

15. Larsen, F., Gundersen, G., Lopez, R., and Prydz, H. (1992) CpG islands as gene markers in the human genome. *Genomics* **13,** 1095–1107.
16. Sambrook, J., Fritsch, E. F., and Maniatis, T., eds. (1990) *Molecular Cloning. A Laboratory Manual.* Cold Spring Harbor Laboratory Press, Cold Spring Harbor, NY.
17. Kuick, R. D., Skolnick, M. M., Hanash, S., and Neel, J.V. (1991) A two-dimensional electrophoresis-related laboratory information processing system: spot matching. *Electrophoresis* **12,** 736–746.
18. Sugahara, Y., Akiyoshi, S., Okazaki, Y., Hayashizaki, Y., and Tanihata, I. (1998) An automatic image analysis system for RLGS films. *Mammalian Genome* **9,** 643–651.
19. Hayashizaki, Y., Hirotsune, S., Hatada, I., Tamatsukuri, S., Miyamoto, C., Furuichi, Y., and Mukai, T. (1992) A new method for constructing NotI linking and boundary libraries using a restriction trapper. *Genomics* **14,** 733–739.
20. Suzuki, H., Kawai, J., Taga, C., Ozawa, N., and Watanabe, S. (1994) A PCR-mediated method for cloning spot DNA on restriction landmark genomic scanning (RLGS) gel. *DNA Res.* **1,** 245–250.
21. Hirotsune, S., Takahara, T., Sasaki, N., Imoto, H., Okazaki, Y., Eki, T., et al. (1996) Construction of high-resolution physical maps from yeast artificial chromosomes using restriction landmark genomic scanning (RLGS). *Genomics* **37,** 87–95.

7

Combined Bisulfite Restriction Analysis (COBRA)

Cindy A. Eads and Peter W. Laird

1. Introduction
1.1. Background

Most molecular biological techniques used to analyze specific loci in complex genomic DNA involve some form of sequence-specific amplification, whether it is biological amplification by cloning in *Escherichia coli*, direct amplification by polymerase chain reaction (PCR), or signal amplification by hybridization with a probe that can be visualized. Since DNA methylation is added postreplicatively by a dedicated maintenance DNA methyltransferase that is not present in either *E. coli* or in the PCR reaction, the methylation information is lost during molecular cloning or PCR amplification. Molecular hybridization does not discriminate between methylated and unmethylated DNA, since the methyl group on the cytosine does not participate in base pairing. The lack of a facile way to amplify the methylation information in complex genomic DNA has been a significant impediment to DNA methylation research.

The indirect methods that have been developed in the past decade to detect DNA methylation patterns at specific loci rely on techniques that alter the genomic DNA in a methylation-dependent manner before the amplification event. There are two main methods that have been utilized to achieve this methylation-dependent DNA alteration. The first is digestion by a restriction enzyme that is affected in its activity by 5-methylcytosine in a CpG sequence context. The cleavage or lack of it can subsequently be revealed by Southern blotting or by PCR. The other technique that has received recent widespread use is the treatment of genomic DNA with sodium bisulfite *(1)*. This treatment converts all unmethylated cytosines in the DNA to uracil by deamination, but leaves the methylated cytosine residues intact. Subsequent PCR amplification replaces the uracil residues with thymines and the 5-methylcytosine residues with cytosines. The resulting sequence difference can be detected using a variety of methods (2).

1.2. Sodium Bisulfite Techniques

Currently, all bisulfite-based methods are followed by a PCR reaction to analyze specific loci within the genome. There are two ways in which the sequence difference generated by the sodium bisulfite treatment can be revealed. The first is to design PCR primers that uniquely anneal with either methylated or unmethylated converted DNA. This technique is referred to as "methylation-specific PCR" or "MSP" (3). The method used by all other bisulfite-based techniques (such as bisulfite genomic sequencing, combined bisulfite restriction analysis [COBRA], and Ms-SNuPE) is to amplify the bisulfite-converted DNA using primers that anneal at locations that lack CpG dinucleotides in the original genomic sequence. In this way, the PCR primers can amplify the sequence in between the two primers, regardless of the DNA methylation status of that sequence in the original genomic DNA. This results in a pool of different PCR products, all with the same length and differing in their sequence only at the sites of potential DNA methylation at CpGs located in between the two primers. The difference between these methods of processing the bisulfite-converted sequence is that in MSP, the methylation information is derived from the presence or absence of a PCR product, whereas in the other techniques a mix of products is always generated and the mixture is subsequently analyzed to yield quantitative information on the relative occurrence of the different methylation states. The best way to view this application is to envision that genomic DNA usually consists of a collection of many different methylation patterns. The PCR reaction will amplify each of these variants without affecting the relative ratio between them. Herein lies both the power and the achilles heel of these techniques. Since all methylation profiles are amplified equally, quantitative information about DNA methylation patterns can be distilled from the resulting PCR pool. However, it is then essential that no bias occurs during the PCR reaction. If a particular sequence variant amplifies with different kinetics than another sequence variant, then incorrect quantitative data will result. Several methods of reducing or preventing PCR bias have been developed (4). An additional difficulty with all non-MSP variants of bisulfite-based DNA methylation analysis techniques is that it is not always easy to find suitable primers lacking CpGs in very dense CpG islands.

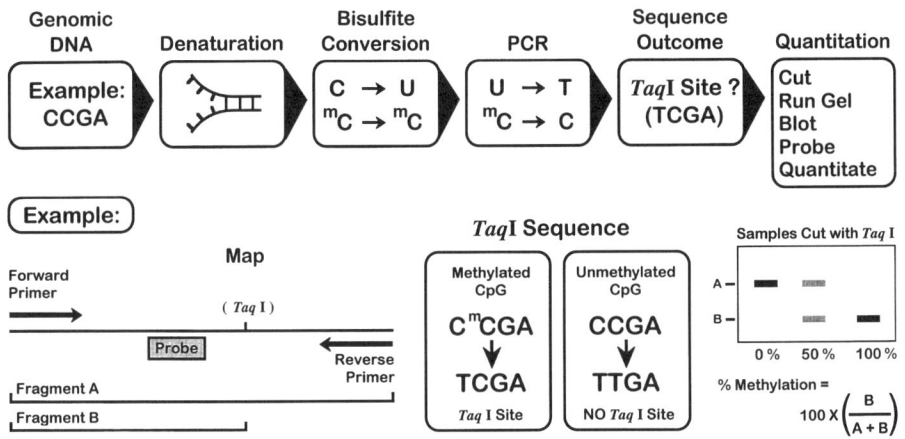

Fig. 1. Outline of the COBRA protocol. The COBRA (COmbined Bisulfite Restriction Analysis) technique relies on the use of restriction-enzyme digestion to detect sequence variations in bisulfite-PCR products caused by differences in DNA methylation patterns in the original genomic DNA. The steps of the technique are illustrated in the boxes at the top. The DNA is denatured and treated with sodium bisulfite, which converts unmethylated cytosine residues to uracils, while methylated cytosines remain unconverted. During PCR, the uracil residues are replaced by thymines, while the methylated cytosine residues are replaced by cytosines. In this example, the consequence is a sequence change that results in the generation of a *Taq*I restriction site only if the cytosine in the CpG dinucleotide is methylated. If the cytosine is unmethylated, then it is converted to thymine and the *Taq*I site is not created. The PCR product is digested by *Taq*I and separated by polyacrylmide gel electrophoresis. The gel is transferred to a membrane and probed by an oligonucleotide for quantitation analysis. The percentage of methylated *Taq*I sites in the target DNA is calculated from the ratio of *Taq*I cleaved PCR product and the total amount of PCR product.

1.3. COBRA

Assuming that the PCR product is a faithful representation of the original collection of DNA sequences following sodium bisulfite treatment, then this resulting pool of PCR products can be analyzed by any technique capable of detecting sequence differences, preferably in a quantitative fashion. COBRA, which stands for COmbined Bisulfite Restriction Analysis, is based on the restriction digestion of the PCR product with an enzyme for which the recognition sequence is affected by the methylation state in the original DNA. Accurate quantitation of the percent methylation can be obtained by subsequent quantitative hybridization. The overall technique is outlined in **Fig. 1**. COBRA can be applied to DNA samples derived from cell lines and tissues. It has even been shown to be compatible with the analysis of microdissected paraffin-embedded sections *(5)*. In addition, COBRA is not as labor-intensive as some of the other bisulfite-based techniques. One interesting feature that COBRA shares with bisulfite genomic sequencing of subclones is that linkage of methylation patterns within individual molecules can be investigated. If two individual CpG sites are interrogated in a COBRA analysis, then a lack of cutting at both sites or complete cutting at both sites with little if any single cutting would suggest that the genomic DNA is comprised of a pool of fully methylated and fully unmethylated DNA molecules and not of mixed methylation. MS-SNuPE interrogates each CpG site independently of other CpGs and would not reveal this distinction.

2. COBRA
2.1. Experimental Design
2.1.1. Sequence Manipulation

Sodium bisulfite treatment substantially alters the primary sequence of genomic DNA. Unmethylated cytosines, which comprise the vast majority of cytosine residues in vertebrate DNA, are chemically converted to uracil residues (which are later substituted by thymine residues during PCR amplification) *(1)*. The immediate consequence of this conversion is that the original DNA strands are no longer complementary. Throughout the genome, the two DNA strands now represent two different, noncomplementary, single-stranded sequences. To design a COBRA assay for a particular region of interest, the first step is the selection of one of the two DNA strand sequences for PCR amplification. For COBRA, the predominant issue in this choice of DNA strand is the location of suitable restriction sites in the final PCR products. One of the confusing concepts for researchers new to bisulfite-based methods is the fact that the amplification of the two opposing strands requires different PCR primers and that the resulting restriction maps of the PCR products will differ. Even the G+C content can differ between the top-strand amplicons and the bottom-strand amplicons, if the distribution of G residues was unequal in the original DNA. To facilitate the design of PCR primers and the localization of restriction-enzyme sites in the subsequent PCR products, it is advisable to perform a theoretical bisulfite conversion for both a fully methylated and completely unmethylated version of both the top and the bottom strand of the sequence of interest. This will generate four new sequences, methylated and unmethylated for both top and bottom, respectively. **Figure 2** illustrates for a sequence example how the methylation status and the strand choice can affect the sequence in the

COBRA

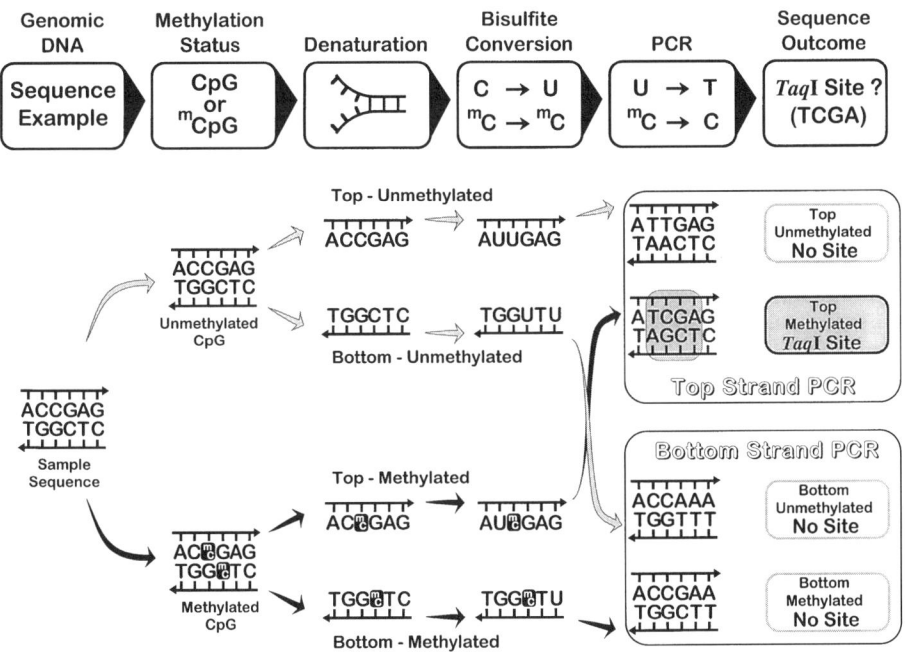

Fig. 2. Sequence outcome and strand choice following bisulfite treatment. Sodium bisulfite modification alters the sequence of DNA such that the top and bottom strands are no longer complementary. For each particular CpG dinucleotide, this results in four new sequence variants, top and bottom unmethylated, and top and bottom methylated. Due to lack of sequence complementarity, the top and bottom strands are amplified in separate PCR reactions using distinct primers designed for either the top or the bottom strands. In COBRA, the PCR primers amplify the the chosen strand independently of methylation status in the original genomic DNA. In this figure, we provide a sequence example to illustrate the complexities of noncomplementarity and methylation status. In our sequence example, amplification of the bottom strand is noninformative, while amplification of the top strand generates a *Taq*I site only if that particular CpG was methylated in the original genomic DNA.

final PCR products. **Figure 3** provides a flow chart for the generation of the four sequence variants.

2.1.2. PCR Primer Design

Once the four sequences and their accompanying restriction maps have been generated, primers can be chosen that flank restriction sites that are differentially present based on the methylation status of the original genomic DNA. In the primer design, it is best to use the bisulfite converted methylated

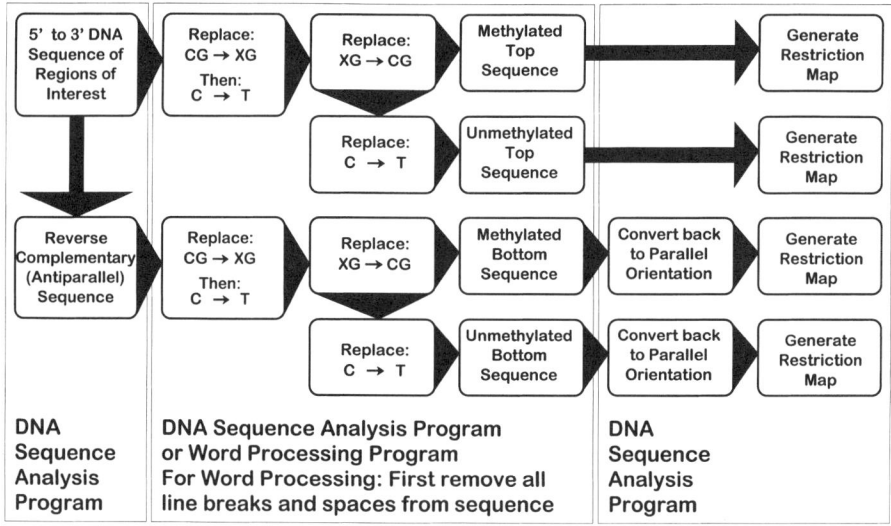

Fig. 3. Computer-simulated bisulfite conversion. In order to design oligonucleotide PCR primers and COBRA probes and to identify restriction-enzyme sites within the PCR product, it is useful to create a theoretical bisulfite conversion for a fully methylated and completely unmethylated form of both the top and bottom strand. This is readily achieved on a computer with a DNA sequence-analysis program. Some of the steps can be performed with a word processing program. Either program can replace all cytosine residues by thymines. Potentially methylated cytosines within a CG sequence context can be masked from this operation by first converting all CG dinucleotides to a neutral unrecognizable sequence such as XG. After the C to T conversion, this neutral sequence can then be restored to CG. It is important to ensure that hidden characters such as spaces and paragraph marks do not interfere with the recognition of CG dinucleotide residues in word processing programs. The bottom methylated and unmethylated versions are developed similarly to the top-strand versions, but the initial DNA sequence is first converted to its antiparallel form. After all the replacements are completed the sequence is changed back to the parallel orientation. In a DNA sequence-analysis program a restriction-enzyme map for each sequence can be generated and CpG dinucleotides highlighted.

sequence versions (top or bottom) as a guide. This facilitates the avoidance of CpG dinucleotides within the primer sequences. It is essential that neither the PCR primers, nor the hybridization probes encompass CpG dinucleotides to ensure equal recognition of DNA regardless of original methylation status. It is helpful to highlight the CpG dinucleotides in the sequence. This can be done by hand or by instructing the DNA analysis software to recognize CpG as a restriction site.

PCR amplification of bisulfite-treated DNA is more difficult than of native DNA. Due to the depletion of cytosine residues from the genomic DNA, the resulting PCR products contain an unequal distribution of bases. Each PCR product contains one C-poor/T-rich strand and one G-poor/A-rich strand. This reduced sequence complexity diminishes the discriminatory ability of the PCR primers and of the hybridization probes. Therefore, it is advisable to design primers slightly longer than for standard PCR. We prefer primers of at least 24 bases. It may be difficult to design such long primers in very dense CpG islands while simultaneously avoiding the inclusion of CpGs within the primer sequences. If possible, other standard criteria for PCR primer design should be adhered to. These include a primer G+C content of 40–60%, similar Tm values for the primer pairs and the avoidance of palindromic or repetitive sequences within the primers and of 3' complementary nucleotides between primer pairs to prevent primer dimer formation.

Due to the difficulties of PCR amplification of bisulfite-treated DNA, it is advisable to design relatively small amplicons. A maximum length of 150 bp is advised if paraffin-embedded samples are to be analyzed. In addition, PCR bias seems to occur less frequently for small amplicons.

2.1.3. Restriction-Enzyme Choice

Some restriction-enzyme sequences are innate in the initial sequence and retained after bisulfite treatment. However, new sites may be generated by the bisulfite conversion and subsequent PCR amplification (*6*). For example, the restriction-enzyme site for *Taq*I (TCGA) can be retained in a methylation-dependent manner or it can be generated by the bisulfite conversion as illustrated in **Figs. 1** and **2**. Sites that are created, rather than merely retained, are preferable, since the use of these sites helps to verify complete bisulfite modification of the DNA. The site will not be created if the bisulfite treatment is insufficient. It is important to stress that the restriction-enzyme cleavage itself is not methylation-dependent. PCR products do not contain 5-methylcytosine. The methylation status is revealed by the presence or absence of a restriction-enzyme site, not by inhibition of cleavage by methylation of the restriction site.

In order to analyze a specific region there must be at least one restriction site within the methylated bisulfite converted strand that is absent in the unmethylated bisulfite-converted strand or vice versa. The easiest way to identify suitable restriction-enzyme sites is to use a DNA-analysis program to generate restriction maps for bisulfite-converted sequences representing the methylated and unmethylated versions of a sequence as described in **Fig. 3**. An advantage of COBRA is that more than one restriction site can be tested on one PCR product given that additional sites are available. A single PCR

**Table 1
Restriction Enzymes Commonly Used in COBRA, with Their Recognition Sequences**[a]

Enzyme	Recognition Site
BsiWI	C/GTACG
BspDI	AT/CGAT
BstBI	TT/CGAA
BstUI	CG/CG
ClaI	AT/CGAT
MluI	A/CGCGT
NruI	TCG/CGA
PvuI	CGAT/CG
TaqI	T/CGA

[a]A unique feature of these restriction enzymes is that their recognition sequences contain cytosines only in the context of CpG.

amplification reaction can be analyzed for any number of restriction enzymes and hybridization probes. **Table 1** shows a list of restriction enzymes commonly used in COBRA analyses.

2.1.4. Probe Design

The design of oligonucleotide probes is facilitated by the restriction maps generated as described in **Fig. 3**. Oligonucleotide probes should not cover either restriction enzyme-recognition sites of the enzymes used in the COBRA analysis, nor should they contain CpG dinucleotide sequences. Longer oligos are easier to use in hybridization reactions, although CpG-rich CpG islands sometimes necessitate the use of probes as short as 15 bases to avoid inclusion of a CpG dinucleotide within the probe.

2.2. Sodium Bisulfite Treatment

The sodium bisulfite conversion of cytosine *(1,7)* proceeds through several steps. Sulfonation of cytosine at the C-6 position *(8)* can only occur on single-stranded DNA. Therefore, it is essential that the genomic DNA is fully denatured and remains denatured until sulfonation is complete. Bisulfite-induced deamination of both methylated and unmethylated cytosine residues occurs *(7)*, but the reactivity of 5-methylcytosine is much lower than that of unmethylated cytosine residues. A competing reaction is the depurination of DNA, which can lead to severe degradation to the point of failure of the PCR reaction *(9)*. The difficulty of sodium bisulfite conversion of genomic

DNA is to find the best balance of complete denaturation of the DNA with complete conversion of unmethylated cytosine residues with minimal DNA loss, depurination and conversion of 5-methylcytosine residues. Various improvements and modifications of the original protocol have been proposed in an attempt to achieve the best balance *(2,9–11)*.

There has been some confusion about concentrations of sodium bisulfite solutions. Sodium bisulfite is sold as a mixture of sodium bisulfite ($NaHSO_3$, FW = 104) and sodium metabisulfite ($Na_2S_2O_5$, FW = 190) (e.g., Sigma S-8890). The ratio can be calculated from the SO_2 assay listed on the bottle. The bottle usually contains mostly sodium metabisulfite. However, since sodium metabisulfite is converted to sodium bisulfite upon dissolution in water (one mole of sodium metabisulfite gives rise to two moles of sodium bisulfite), it is simpler to purchase pure (98.8%) sodium metabisulfite (Sigma S-1516) and dissolve it at half the specified sodium bisulfite concentration. For example, for a 5 M concentration of sodium bisulfite, prepare a 2.5 M solution of sodium metabisulfite. The preparation of high concentrations of sodium (meta)bisulfite require the addition of NaOH to allow the salt to go into solution.

Sodium bisulfite oxidizes easily. Therefore, care should be taken not to excessively aerate during the dissolving process. Heating to 50°C may be necessary to obtain complete dissolution of the sodium bisulfite. Hydroquinone is added to prevent oxidation of the sodium bisulfite. The original protocol calls for a freshly prepared 3.6 M solution of sodium bisulfite to be added, along with the hydroquinone to the denatured DNA to give a final concentration of 3.1 M sodium bisulfite, followed by a 16-h incubation at 55°C *(1)*. However, variations in the concentration of sodium bisulfite, the incubation temperature and time may yield a better balance of complete conversion, while maintaining the integrity of the DNA *(9)*. Also, it is more efficient to combine the sodium bisulfite and hydroquinone solutions before adding them to the individual denatured DNA samples. The procedure described in **Subheading 3.2.** is a modification of the original protocol *(1)*. We recommend a 4.75 M sodium bisulfite treatment for 12–16 h at 50°C. However, investigators are encouraged to experiment with time, temperature, and sodium bisulfite concentrations to find the best conditions for their particular application.

Incomplete conversion of unmethylated cytosine residues is occasionally seen. Therefore, it is essential to check this for the conditions used in a particular application. Complete conversion of the DNA can be readily verified by restriction digestion with an enzyme that contains a cytosine in the recognition sequence that is not within a CpG sequence context *(5)*. Such sites should be completely lost during bisulfite conversion since the unmethylated cytosine should be converted to thymine. Any cutting of the PCR product by such an enzyme indicates either non-CpG methylation or incomplete bisulfite conversion. Comparison of the restriction maps of the unconverted sequence with the maps generated according to **Fig. 3** should yield several choices of control enzymes. The addition of urea can improve the efficiency of conversion by maintaining the DNA in a denatured state *(11)*.

2.3. PCR

As with any PCR reaction, initial optimization of the thermal-cycling parameters is advisable. The lower sequence complexity of the bisulfite converted DNA and the amplification primers and potential degradation of the DNA by depurination contribute to the difficulty of bisulfite PCR reactions. Initial denaturation of the DNA for 2–4 min at 95°C in the first cycle seems to be beneficial. A 1-min denaturation can suffice for subsequent cycles. We obtain better results with mixtures of Taq polymerase and high-fidelity polymerases such as Roche Boehringer Mannheim Expand Hi Fidelity polymerase. The number of cycles needed to generate a product depends on the number of starting molecules. For cell-line and tissue DNA samples the amount of DNA is often at the microgram level, in which case 30 cycles are more than sufficient to generate a robust PCR product. However, since paraffin-embedded samples may have less than a nanogram of DNA initially and subsequent loss and degradation of DNA occurs during the bisulfite treatment, it may be necessary to increase the number of cycles to 40. In extreme cases, nested PCR can be employed, but this increases the risk of PCR bias. Other parameters, such as $MgCl_2$ and primer concentrations should be optimized as for any PCR assay.

2.4. Restriction Enzyme Digestion

Following PCR amplification the product must be cleaned up before further restriction-digestion analysis. The residual salts from the PCR buffers may inhibit complete enzyme digestion. In addition, some proprietary PCR buffers, such as those that are supplied with the Expand polymerase, contain components that are inhibitory to restriction-enzyme digestion. If the PCR produces a strong single band on an agarose gel, then the product can be simply purified by a commercial PCR clean-up kit or microfiltration spin column. However, if nonspecific PCR products result from amplification, then gel extraction of the desired product is recommended. Restriction digestion is performed according to manufacturers specifications.

2.5. Polyacrylamide Gel Electrophoresis, Electroblotting, and Hybridization

After restriction digestion of the purified PCR product, the sample is separated on an 8% denaturing polyacrylamide gel. The large volume of the restriction digestion can be a problem in achieving a fourfold volume of denaturing loading dye, which is essential to prevent secondary structures. A high concentration of pure PCR product will allow the use of fairly small restriction-digestion volumes. In addition, the use of small protein gels with thick spacers allows for larger loading volumes and provides sufficient separation of digestion products. The extra thickness of the gels also facilitates manipulation of the gel during the electroblot set up. High-concentration agarose gels do not blot efficiently.

Once the gel electrophoresis is completed, the DNA is transferred to a positively charged membrane (Zetabind Membrane, American Bioanalytical) by electroblotting. Electroblotting provides a fast and efficient transfer of the DNA. Following electroblotting, the DNA is crosslinked to the membrane by exposing the membrane to 1200J of UV. At this stage the membrane can be processed immediately or wrapped in Saran wrap and stored at 4°C for future use.

The membrane is prepared for hybridization by first prewetting it in 6X SSC and then soaking it in prehybridization buffer at 42°C for at least 30 min. The membranes are then hybridized with a 5'-end labeled oligonucleotide (15–25 nucleotides) overnight. The hybridization reaction depends on the length, concentration, base composition, and Tm of the probe. In general, the hybridization should be performed at 5°C below the Tm of the probe. However, if the probe is less than 15–20 nucleotides, then less stringent conditions are recommended. Specialized hybridization buffers and conditions that have been optimized for oligo hybridization can also be employed.

For short probes (less than 15–20 nucleotides), the membrane is washed at low stringency using 2X SSC/1% SDS at 25–37°C. Longer probes allow more stringent conditions, 0.5–2X SSC/1% SDS at 42–65°C. The membranes are washed in a volume of 200 mL with replacement of the washing solution at 10–15 min intervals. Oligo probes do not require very long washing. We usually wash less than an hour.

The filter is then exposed in a phosphoimager system which allows quantitation of the individual bands. DNA methylation levels are calculated according to the formula shown in **Fig. 1**.

3. Methods
3.1. Materials
3.1.1. DNA Isolation

1. Phosphate-buffered saline (PBS).
2. Lysis Buffer: 100 mM Tris-HCl, pH 8.5, 10 mM ethylenediaminetetraacetic acid (EDTA), 200 mM NaCl, 1% SDS.
3. Proteinase K.

4. RNase A.
5. Phenol/chloroform/isoamyl (25:24:1).
6. Ethanol.
7. TE Buffer: 10 mM Tris-HCl, pH 7.5, 0.1 mM EDTA.
8. For Paraffin-embedded sections: Buffer K: 10 mM Tris-HCl, pH 8.0, 5 mM EDTA, 10 µg/µL Proteinase K.

3.1.2. Bisulfite Treatment

1. 1–10 µg DNA.
2. Salmon sperm DNA (only if initial DNA is < 1 µg).
3. 2 M NaOH, a 5 M sodium bisulfite, pH 5.0, made from sodium metabisulfite (Sigma S-1516), which also contains 125 mM Hydroquinone (*see* **Subheadings 2.2. and 3.2.2**).
4. Mineral oil.
5. Wizard DNA clean-up System (Promega).
6. 3cc syringes.
7. 80% isopropanol.
8. 3 M NaOH
9. 5 M Ammonium Acetate, pH 7.4).
10. Ethanol (100% and 70%).
11. TE Buffer: 10 mM Tris-HCl, pH 7.5, 0.1 mM EDTA.

3.1.3. PCR

1. 10X Expand buffer (Boehringer Mannheim).
2. 25 mM Expand MgCl (Boehringer Mannheim).
3. 4 mM dNTPs.
4. 50 µM of each primer.
5. 2 units of Expand HF Enzyme (Boehringer Mannheim).
6. Sterile ddH$_2$0.

3.1.4. DNA Extraction

1. Commercial PCR clean-up kit or gel extraction kit.

3.1.5. Restriction Enzyme Digestion

1. 10 ng of purified PCR product.
2. 1 unit of enzyme.
3. 1X enzyme buffer
4. Sterile ddH$_2$O.
5. Mineral oil.

3.1.6. Polyacrylamide Gel Electrophoresis

1. 8% denaturing polyacrylamide gel (7 M Urea).
2. 10% Ammonium persulfate.

3. TEMED.
4. Loading buffer: 98% formamide, 10 mM EDTA, 0.0025% xylene cyanol, 0.0025% bromophenol blue.
5. 1X TBE Buffer.

3.1.7. Electroblotting

1. Zetabind charged membrane (American Bioanalytical).
2. Blotting paper.
3. Fiber pads (Bio-Rad).
4. 0.5X TBE.

3.1.8. Hybridization

1. End-labeling probe: 10 pmoles oligonucleotide, 10X T4 polynucleotide kinase buffer, 20 units T4 polynucleotide kinase, γ-^{32}P-ATP (10 mCi/mL), sterile ddH$_2$O, G-50 sephadex, 1 cc syringe.
2. Prehybridization: 6X SSC, 10 mL hybridization buffer: 500 mM phosphate buffer, pH 6.8, 1 mM EDTA, 7% SDS *(12)*.

3.2. Methods

3.2.1. DNA Isolation

Genomic DNA can be collected from cultured cells, tissue, or paraffin-embedded sections by standard techniques *(13,14)*. The quality of the DNA retrieved from this isolation procedure is appropriate for COBRA analysis. It is important to remove all traces of RNA from the sample in order to prevent RNA amplification during PCR.

3.2.2. Bisulfite Treatment

Carrier DNA needs to be added to samples that contain less than 1 µg of DNA, since there can be substantial loss and degradation of DNA during the bisulfite conversion. We use 1–2 µg of salmon sperm DNA as a carrier. Embedding the DNA in agarose beads provides a good alternative to prevent loss of the DNA during the procedure *(10)*.

1. Add 1–10 µg of DNA to sterile ddH$_2$O for a final volume of 18 µL and heat at 95°C for 20 min. Remove the sample from the heat and immediately place on ice.
2. Add 2 µL of 3 M NaOH to denature the DNA and incubate for 20 min at 42°C.
3. Prepare a fresh solution of 5 M sodium bisulfite solution/hydroquinone solution, pH 5.0, immediately prior to use. Calculate 0.5 mL of solution for each DNA sample. For 4 mL of solution (8 samples), add 1.9 g sodium metabisulfite to 2.5 mL sterile ddH$_2$O. Add 0.7 mL 2 M NaOH. Add 0.5 mL 1 M hydroquinone (0.11 g in 1 mL H$_2$O). Heat to 50°C. Invert frequently until fully dissolved.

4. Add 380 μL of the 5 M sodium bisulfite/hydroquinone solution to the 20 μL of denatured DNA. Mix.
5. Overlay sample with 5–6 drops of mineral oil to prevent evaporation and incubate 12–16 h at 50°C in the dark. Significant degradation may occur at longer incubation times.
6. Desalt the sample with Wizard DNA clean-up system.
7. Transfer to a clean tube and elute the DNA with 45 μL of sterile ddH$_2$O.
8. Add 5 μL of 3 M NaOH and incubate for 15 min at 37°C for desulphonation.
9. Add 75 μL of 5 M Ammonium Acetate pH 7.4. Mix.
10. Add 2.5 volumes of 100% ethanol and precipitate for 0.5–1 h at –70°C.
11. Spin at maximum speed in microcentrifuge for 20 min.
12. Wash DNA pellet with 70% ethanol and resuspend DNA in 30–50 μL of TE, pH 7.5.
13. Store bisulfite samples in the –20°C freezer. The life-span of bisulfite treated DNAs can exceed 12 mo when stored properly.

3.2.3. PCR Amplification

PCR amplification is performed using standard PCR protocols (*see* **Subheading 2.3.**).

3.2.4. DNA Clean Up

Clean PCR product as described in **Subheading 2.4.**

3.2.5. Restriction-Enzyme Digestion

1. Digest 10 ng of purified PCR product with 1 U of enzyme in a 10–15 μL reaction for at least 4 h to ensure complete digestion. See **Subheading 2.1.3.** for guidance on restriction-enzyme choice.
2. Overlay the sample with mineral oil to prevent evaporation.

3.2.6. Polyacrylamide Gel Electrophoresis

1. Add 4 volumes of loading buffer to each digested sample. The mineral oil does not need to be removed.
2. Denature for 2–3 min at 95°C.
3. Load sample onto a 55°C prewarmed 8% denaturing polyacrylamide gel (7 M urea).
4. Run gel in 1X TBE at 50–60 milliAmps.

3.2.7. Electroblotting, Prehybridization, Hybridization, and Washing

1. Transfer the DNA to Zetabind Membrane by electroblotting for at least 1 h at 20 volts.
2. UV crosslink the DNA to the membrane at 1200 Joules.
3. Prewet the nylon membrane in 6X SSC and place membrane into a hybridization bottle.

4. Add 10 mL of Church prehybridization buffer and rotate for 0.5–1 h at 42°C.
5. Add 5′ end-labeled probe (10 pmoles) and hybridize overnight at appropriate temperature.
6. Wash the membrane as described in **Subheading 2.5.**

References

1. Frommer, M., McDonald, L. E., Millar, D. S., Collis, C. M., Watt, F., Grigg, G. W., et al. (1992) A genomic sequencing protocol that yields a positive display of 5-methylcytosine residues in individual DNA strands. *Proc. Natl. Acad. Sci. USA* **89,** 1827–1831.
2. Rein, T., DePamphilis, M. L., and Zorbas, H. (1998) Identifying 5-methylcytosine and related modifications in DNA genomes. *Nucleic Acids Res.* **26,** 2255–2264.
3. Herman, J. G., Graff, J. R., Myöhänen, S., Nelkin, B. D., and Baylin, S. B. (1996) Methylation-specific PCR: a novel PCR assay for methylation status of CpG Islands. *Proc. Natl. Acad. Sci. USA* **93,** 9821–9826.
4. Warnecke, P. M., Stirzaker, C., Melki, J. R., Millar, D. S., Paul, C. L., and Clark, S. J. (1997) Detection and measurement of PCR bias in quantitative methylation analysis of bisulphite-treated DNA. *Nucleic Acids Res.* **25,** 4422–4426.
5. Xiong, Z. and Laird, P. W. (1997) COBRA: a sensitive and quantitative DNA methylation assay. *Nucleic Acids Res.* **25,** 2532–2534.
6. Sadri, R. and Hornsby, P. J. (1996) Rapid analysis of DNA methylation using new restriction enzyme sites created by bisulfite modification. *Nucleic Acids Res.* **24,** 5058–5059.
7. Wang, R. Y., Gehrke, C. W., and Ehrlich, M. (1980) Comparison of bisulfite modification of 5-methyldeoxycytidine and deoxycytidine residues. *Nucleic Acids Res.* **8,** 4777–4790.
8. Hayatsu, H. (1976) Bisulfite modification of nucleic acids and their constituents. *Prog. Nucleic Acid Res. Mol. Biol.* **16,** 75–124.
9. Raizis, A. M., Schmitt, F., and Jost, J. P. (1995) A bisulfite method of 5-methylcytosine mapping that minimizes template degradation. *Anal. Biochem.* **226,** 161–166.
10. Olek, A., Oswald, J., and Walter, J. (1996) A modified and improved method for bisulphite based cytosine methylation analysis. *Nucleic Acids Res.* **24,** 5064–5066.
11. Paulin, R., Grigg, G. W., Davey, M. W., and Piper, A. A. (1998) Urea improves efficiency of bisulphite-mediated sequencing of 5′- methylcytosine in genomic DNA. *Nucleic Acids Res.* **26,** 5009–5010.
12. Church, G. M. and Gilbert, W. (1984) Genomic sequencing. *Proc. Natl. Acad. Sci. USA* **81,** 1991–1995.
13. Laird, P. W., Zijderveld, A., Linders, K., Rudnicki, M. A., Jaenisch, R., and Berns, A. (1991) Simplified mammalian DNA isolation procedure. *Nucleic Acids Res.* **19,** 4293.
14. Shibata, D. (1992) The polymerase chain reaction and the molecular genetic analysis of tissue biopsies, in *Diagnostic Molecular Pathology: A Practical Approach, vol. II* (Herrington, C. S. and McGee, J. O. D. eds.), IRL Press, Oxford, UK, pp. 85–111.

8

Differential Methylation Hybridization Using CpG Island Arrays

Pearlly S. Yan, Susan H. Wei, and Tim Hui-Ming Huang

1. Introduction

Differential hybridization and its related techniques have been developed to identify genes whose expressions are altered during different physiological conditions or in dissimilar cell or tissue types *(1–3)*. High-throughput DNA array technologies have further advanced these analyses, providing for simultaneous examinations of expressions of thousands of genes between two different cell types in a single experiment *(4–6)*. Recently, we have adapted the hybridization approach to devise an array-based technique, called differential methylation hybridization (DMH), to identify changes in DNA methylation patterns commonly observed in cancer *(7)*. Methylation of DNA is the addition of a methyl group to the 5th carbon position of cytosine that is 5' to a guanine in GC-rich regions known as CpG islands, which are frequently located in the 5'-end of regulatory regions of genes *(8,9)* This epigenetic reaction does not usually change nucleotide sequences nor affect the specificity of DNA base pairing, but it can have inhibitory effects on gene expression *(8,9)*.

The DMH methodology in essence comprises of three fundamental components: the panel array of CpG island clones, the DNA samples under investigation, and hybridization of the sample DNAs onto the CpG arrays. Unlike cDNA arrays that use expressed sequence tags as hybridization templates, the DMH approach uses genomic DNA sequences that are GC-rich and contain restriction sites that are sensitive to methylation for the testing. Through the elegant work of Cross et al. *(10)*, a human genomic library called CGI that is specifically enriched for 0.2–2 kb CpG island sequences was made available to us. To generate a membrane panel of CpG islands, CGI clones negative for

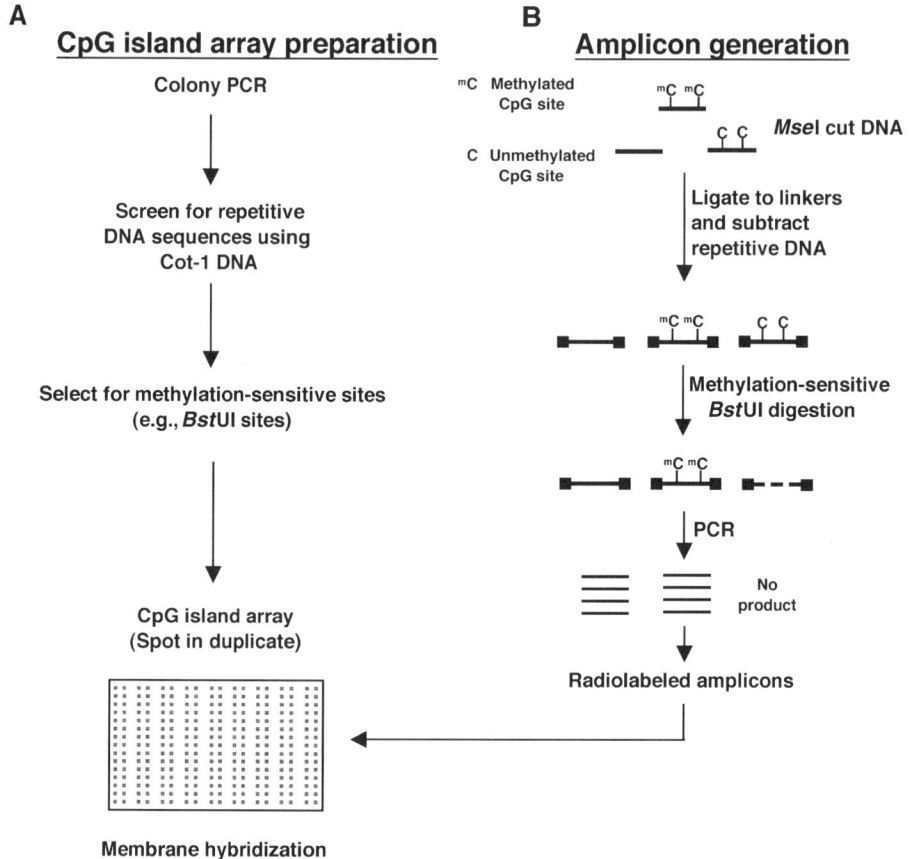

Fig. 1. Schematic flowchart for differential methylation hybridization. (**A**) The selection of CpG island genomic clones gridded on high-density arrays. (**B**) The preparation of amplicons used as hybridization probes.

repetitive sequences were first identified by colony hybridization, then selected for polymerase chain reaction (PCR) amplification and arrayed onto solid supports (**Fig. 1A**). A four-base restriction enzyme, such as *Bst*UI, *Hpa*II, or *Hha*I, whose recognition sites contain the CG dinucleotides and are methylation-sensitive for restriction, is key to the DMH analysis.

The interrogating probes, or "amplicons," can be prepared from the tumor and its corresponding normal sample (**Fig. 1B**). Genomic DNA is first cut with a four-base cutter, *Mse*I, which restricts bulk DNA into smaller fragments of <200 bps but leaves the larger GC-rich CpG island fragments relatively intact *(10)*. This latter fraction is then ligated to linker primers for subsequent PCR amplifications. Repetitive sequences are next depleted from the DNA samples

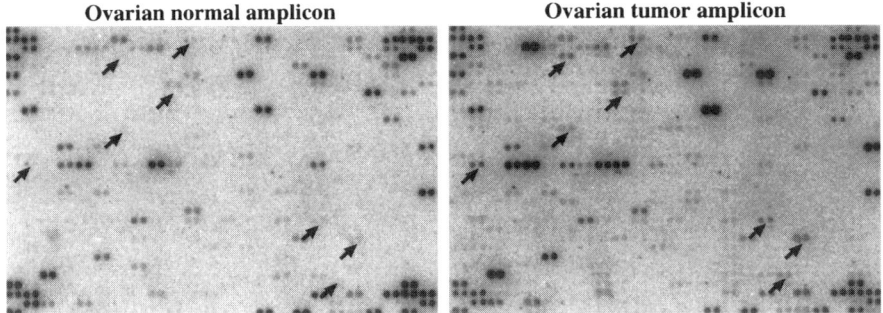

Fig. 2. Representative results of differential methylation hybridization. PCR products of a total of 736 CpG island clones were dotted onto membrane in duplicate and hybridized first with ^{32}P-labeled amplicons from normal ovarian tissue. The same membrane was later hybridized with amplicons derived from ovarian tumor. Eight control DNA samples were dotted in quadruplicate on the four corners of the array to serve as orientation marks and for comparison of hybridization signal intensities. Arrows mark the position of some "hypermethylated" clones, showing greater signal intensities in the panel hybridized with tumor amplicon than in the same panel hybridized with normal amplicon.

by subtractive hybridization using human Cot-1 DNA, and the samples are digested with a methylation-sensitive restriction enzyme. The overall DNA sample is lastly amplified by linker-PCR prior to hybridization onto the membrane. In this design, genomic fragments containing unmethylated CG sites in one DNA sample, e.g., normal control, are cut and cannot be amplified, whereas these same fragments containing these restriction sites that are potentially methylated in the tumor sample are protected from digestion and are subsequently amplified by PCR. The differentially methylated sequences may therefore be identified by comparing the hybridization signal intensities between the tumor and normal amplicons on the CpG island arrays **(Fig. 2)**. CpG islands that are candidates for aberrant methylation in cancer cells can then be cloned for further characterization. In addition to cancers, DMH may be applicable for studying changes of DNA methylation in aging and cell differentiation and development.

2. Materials

1. Human CpG island CGI genomic library (United Kingdom Human Genome Mapping Project Resource Centre; http://www.hgmp.mrc.ac.uk/).
2. 96-well toothpick holder, 96-pin MULTI-BLOT™ replicator system, and 4 × 4 MULTI-PRINT register frame (V&P Scientific, San Diego, CA).
3. 96-well culture plates (Fisher Scientific, St. Louis, MO).

4. SealPlate™ adhesive sealing films for 96-well culture plates (Midwest Scientific, St. Louis, MO).
5. 96-well microplates and Microseal™ 'A' film (MJ Research, Watertown, MA).
6. RAGE® Rapid Agarose Gel Electrophoresis. RGX-100 System (Cascade Biologics, Inc., Portland, OR).
7. PCR primers: HGMP 3558.1: 5'-CGG CCG CCT GCA GGT CTG ACC ATA A-3'; HGMP 3559.1: 5'-AAC GCG TTG GGA GCT CTC CCA TAA-3'; H-12: 5'-TAA TCC CTC GGA-3'; H-24: 5'-AGG CAA CTG TGC TAT CCG AGG GAT-3' (GibcoBRL/Life Technologies, Rockville, MD).
8. 10X TBE buffer, per L: Trizma, 109 g; boric acid, 55 g; 0.5 M Na$_2$EDTA, 40 mL.
9. T4 DNA ligase (400 U/μL supplied with 10X ligase buffer), Deep Vent$_R$® $^{(exo-)}$ DNA polymerase (2 U/μL, supplied with 10X ThermoPol reaction buffer), restriction enzymes MseI (4 U/μL, supplied with 10X NEBuffer 2) and BstUI (10 U/μL, supplied with 10X NEBuffer 2 and 100X BSA at 10 mg/mL from which 10X BSA is prepared), and NEBlot® Kit (supplied with random octadeoxyribonucleotides in 10X labeling buffer, DNA polymerase I-Klenow fragment, 5 U/μL, and dNTPs, 0.5 mM in Tris-HCl, pH 7.0) (New England BioLabs, Beverly, MA).
10. QIAquick PCR purification kit (contains QIAquick spin columns, buffer PB, and buffer PE) (Qiagen, Valencia, CA).
11. Nytran N nylon transfer membrane (Schleicher & Schuell, Keene, NH).
12. QuickDraw™ blotting paper, ampicillin, dimethyl sulfoxide (DMSO), bromphenol blue-xylene cyanole dye solution (Sigma, St. Louis, MO).
13. Deoxynucleotide triphosphates (dNTPs), 100 bp DNA ladder, and nick translation enzymes: DNA polymerase I (10 U/μL, supplied with DNA polymerase I 10X reaction buffer) and RQ1 RNase-free DNase (1 U/μL) (Promega, Madison, WI).
14. 0.4 mM Biotin-14-dCTP, human Cot-1 DNA (1 mg/mL for Cot-1 subtraction and for hybridization subtraction, 10 ng/μL for labeling reactions), 20X SSC buffer, 1 M Tris buffer pH 8.0, 5 M NaCl, and 0.5 M ethylenediaminetetraacetic acid (EDTA) (GibcoBRL/Life Technologies).
15. Streptavidin magnetic particles and G-50 Sephadex® Quick Spin™ columns (Boehringer Mannheim, Indianapolis, IN).
16. Redivue [α]^{32}P-dCTP: 3,000 Ci/mmol (Amersham Pharmacia Biotech, Piscataway, NJ).
17. High-efficiency hybridization solution (HEHS) (Molecular Research Center, Cincinnati, OH).

3. Methods
3.1. CpG Island Array Preparation
3.1.1. Colony PCR

1. *Escherichia coli* harboring plasmids of the CGI library are grown on LB agar plates according to standard procedures.
2. To prepare for amplification of gene inserts, individual colonies are first grown in a 96-well culture plate containing 50 μL LB broth plus ampicillin (50 μg/mL)

Table 1
Master Mix for Colony PCR

Reagents	Per sample (μL)	Per 100 samples (μL)
Water	17.5	1750
10X ThermoPol buffer	2.5	250
DMSO	2.5	250
HGMP 3558.1 (4 μM)	1	100
HGMP 3559.1 (4 μM)	1	100
dNTP (2.5 mM each)	0.25	25
Deep Vent$_R$ ($^{exo-}$) DNA Polymerase (2 U/μL)	0.25	25
Total volume	25	2500

per well. A 96-hole toothpick holder is placed over the plate and autoclaved toothpicks are used to transfer and deposit bacterial cells into each well of the 96-well plate. The culture plate is sealed with SealPlate™ plastic film and the colonies are grown in a 37°C shaker incubator overnight.

3. In order to amplify the inserts in a 96-well microplate, a master mix is prepared and 25 μL is pipetted to each well (*see* **Table 1**).
4. The 96-pin MULTI-BLOT replicator is dipped into the culture microplate, transferred to the 96-well plate containing the PCR reaction, and swirled to mix. Quickly re-dip the replicator into the culture plate a second time and touch the PCR mixture again but without swirling. Seal the PCR microplate with Microseal 'A' Film.
5. The following PCR program (PTC-100 thermocycler, MJ Research) is utilized for the amplification:
   Step 1: 98°C for 4 min → Step 2: 95°C for 45 s → Step 3: 55°C for 45 s
   Step 4: 72°C for 1 min → Step 5: Go to Step 2 for 29 times →
   Step 6: 72°C for 5 min → Step 7: 4°C to ∞.
6. At the completion of the PCR cycles, 1 μL of the PCR product from each well is pipetted to a new 96-well microplate for determining the presence of methylation-sensitive sites using a restriction enzyme. To the remainder of the PCR sample, 2 μL bromophenol blue-xylene cyanol tracking dye is added for using them to screen-repetitive sequences in a Cot-1 hybridization procedure (*see* **Subheading 3.1.3.**).

3.1.2. Methylation-Sensitive Restriction Digest

1. A master mix is prepared for the four-base cutting enzyme *Bst*UI to the quantity of 14 μL per well in the new 96-well microplate that contains 1 μL PCR products (*see* **Table 2**).
2. This 96-well PCR microplate is sealed with Microseal 'A' Film and incubated at 60°C for 1–2 h.

Table 2
Master Mix for Restricting PCR-Amplified Tags

Reagents	Per sample (µL)	Per 100 samples (µL)
Water	12	1200
10X NEB2 Buffer	1.4	140
BstUI (10 U/µL)	0.6	60
Total volume	14	1400

3. At the end of digestion, 2.5 µL tracking dye is added to each sample and analyzed on a 1% agarose gel using the 96-well gel electrophoresis system of RAGE®. Ten microliters (10 µL) of the digest is loaded side by side to 1 µL of the undigested product of each clone for comparison.
4. When electrophoresis is completed, results are documented with a Polaroid film.
5. Evaluate the BstUI digestion patterns. Only clones that give BstUI-digested bands lower than the undigested PCR product (indicating the presence of BstUI sites) are considered for printing onto the panel arrays.

3.1.3. Cot-1 Hybridization

1. The CGI library is further screened for repetitive sequences by blot hybridization using human Cot-1 DNA as the probe. Mount a Nytran membrane onto the 4 × 4 MULTI-PRINT register frame (with 2 sets of 16 alignment holes oriented toward the bottom of the membrane) and support the assembly on a solid smooth surface, such as a glass pane. This register frame will allow the printing of up to 1,536 single spots (from sixteen 96-well plates) or 768 double spots (from eight 96-well plates).
2. To spot the CGI clones onto the membrane, first denature the colony PCR products from the 96-well plate at 95°C for 5 min and then chill on ice. Dip the clean replicator pins into the denatured PCR products with the guard pin oriented to the bottom of the plate. Transfer the PCR products to the membrane by placing the guard pin in the first alignment hole on each side of the register frame. Each pin transfers a ~0.4 µL drop (~40 ng DNA) onto the membrane. Every set of PCR products is printed singly for five times in this manner. The second 96-well PCR plate is printed by using the next alignment hole.
3. Once all the sets of PCR products are printed, the membranes are denatured on blotting paper saturated with, but not immersed in, the denaturing buffer (0.5 M NaOH and 1.5 M NaCl, pH 10.0–14.0) for 1 min. At the end of the denaturation step, the membrane is transferred to a blotting paper saturated with neutralizing buffer (0.5 M Tris-HCl and 1.5 M NaCl, pH 7.0–8.0) for 3 min. The membrane is lastly moved onto a blotting paper saturated with 5X SSC buffer for 3 min. The membrane is then air-dried and cross-linked with UV.

4. Prior to hybridization of the CGI clones with human Cot-1 DNA, the tracking dye is removed from the membrane by washing it in a solution of 0.1% SDS and 0.2X SSC at 55°C for 12–15 min.
5. To label the probe, 165 ng human Cot-1 DNA (10 ng/μL) is first denatured at 95°C for 5 min and placed on ice. The ^{32}P-labeling reaction, the membrane hybridization and the posthybridization washing are carried out as described in **Subheading 3.3.** with the exception that human Cot-1 DNA is not included as a repetitive sequence suppressor.
6. An overnight exposure to X-ray film or a 4-h exposure to a PhosphorImager cassette is sufficient to clearly image Cot-1 positive clones.
7. Thus from the analyses of *Bst*UI digests and Cot-I hybridization of the CGI library, clones that are screened positive for *Bst*UI and negative or weakly positive for human Cot-1 are selected and grown in a new set of 96-well cell-culture plates as before in 100 μL LB per well with ampicillin (50 μg/mL). Colony PCR reactions for these clones are carried out according to **steps 3–5** in **Subheading 3.1.1.** with the exception that the four corners of each PCR plate are reserved for positive internal controls. A test gel is performed again to check for successful PCR amplification as described in **steps 3–4** in **Subheading 3.1.2.** Meanwhile, a –70°C freezer stock of these culture plates is prepared by transferring the selected CGI clones with the 96-pin replicator 10 times to 10 μL LB broth containing ampicillin in 60% glycerol. It is recommended that a second set of CGI panels be prepared and stored in a different –70°C freezer for additional safeguards.
8. The printing of the CpG island panels is essentially the same as previously described in **steps 1–3** of this section. The final volume of 25 μL PCR master mix and 6 μL tracking dye will be sufficient for printing 15–20 panels with each spot printed 5 times in duplicates. After denaturation, neutralization, and UV cross-linking, the nylon membranes is then ready for hybridization with the amplicons or can be stored at –20°C until used.

3.2. Amplicon Generation

3.2.1. MseI Digestion

1. High-molecular-weight DNA (0.2–0.5 μg) is cut with *Mse*I according to the digestion mixture listed in **Table 3**.
2. The digestion mixture is layered with autoclaved mineral oil and placed in a 37°C water bath overnight.
3. The digested DNA is purified using the QIAquick PCR Purification Kit according to the manufacturer's specifications. Briefly, 250 μL PB buffer (or 5X the digestion volume) is added to the *Mse*I-digested DNA and thoroughly mixed prior to applying it to the QIAquick spin column. The column is centrifuged and washed with 750 μL PE buffer. The digested DNA is eluted twice with 50 μL water and vacuum-dried.

Table 3
Mix for Restricting Genomic DNA

Reagents	Per sample (µL)	Per 8 samples (µL)
DNA/water	35	280
10X NEB2 buffer	5	40
10X BSA (1 mg/mL)	5	40
*Mse*I (4 U/µL)	5	40
Total volume	50	400

Table 4
Reaction Mix for Ligating Primer Linkers to Genomic DNA

Reagents	Per sample (µL)	Per 8 samples (µL)
H-12/H-24 (100 µM each)	7	56
10X Ligase buffer	2.5	20
T4 Ligase (400 U/µL)	2	16
Total volume	11.5	92

3.2.2. Linker Ligation

1. The ligation step begins with the addition of 13.5 µL water to the dried DNA samples. Primers H-12 and H-24 (each 100 µM) are first annealed to each other by mixing equal amounts of each in a microcentrifuge tube, and allowing the mixture to cool gradually from 55°C to room temperature. The annealed primers are then added to the DNA. The ligation mixture is as listed in **Table 4** and should be placed on ice at all times.
2. The reaction is carried out in a 14°C water bath for 6 h and then frozen until the sample is ready to be processed further.
3. Optional: To determine the completeness of the *Mse*I digestion and the efficiency of the ligation step, carrying out a test PCR reaction is recommended. One microliter (1 µL) of the ligated DNA is added to 19 µL of the mixture listed in **Table 5**.

    This PCR mixture is layered with autoclaved mineral oil and are amplified as follows (Step 1 functions to fill in the protruding ends of the ligated DNA):

    Step 1: 72°C for 5 min → Step 2: 97°C for 1 min → Step 3: 72°C for 3 min → Step 4: Go to Step 2 for 24 times → Step 5: 72°C for 10 min → Step 6: 4°C to ∞.
4. Ten microliters (10 µL) of the above PCR product is analyzed on a 1% agarose gel alongside 1 µL of 100 bp DNA ladder. A diffuse smear with a strong presence between 0.2–2.0 kb would indicate successful digestion and ligation reactions (*see* **Note 1**).

Table 5
Reaction Mix to Test Efficiency of Primer Ligation

Reagents	Per sample (μL)	Per 8 samples (μL)
Water	13.7	109.6
DMSO	2	16
10X ThermPol buffer	2	16
H-24 (10 μM)	0.5	4
dNTP (10 mM)	0.4	3.2
Deep Vent$_R$ ($^{exo-}$) DNA Polymerase (2 U/μL)	0.4	3.2
Total volume	19	152

Table 6
Labeling Cot-1 DNA with Biotin

Reagents	Per sample (μL)	Per 8 samples (μL)
Water	102	816
10X DNA Polymerase I buffer	20	160
dNTP (–dCTP, 300 μM each)	37.5	300
Human Cot-1 DNA (1 mg/mL)	15	120
Biotin-14-dCTP (0.4 mM)	19.5	156
DNA Polymerase I (10 U/μL)	3.75	30
RQ1 RNase-free DNase (1 U/μL)	2.25	18
Total volume	200	1600

3.2.3. Cot-1 Subtraction

1. To remove the majority of the repetitive elements from the ligated DNA sample, a human Cot-1 DNA subtraction step is performed. Fifteen to twenty micrograms (15–20 μg) of the human Cot-1 DNA (1 mg/mL) is labeled with biotin for each ligated DNA sample by Nick Translation using the master mix in **Table 6**. All reagents should be kept on ice.
2. Incubate the Nick Translation reaction mix in a 14°C water bath for 1.5 h. Freeze the reaction mix at –20°C until subtractive hybridization is ready.
3. Optional: To determine the efficiency of the labeling of Cot-1 with biotin, a "tracer" reaction with ^{32}P is suggested as a test before proceeding further with preparations of the amplicon DNAs. (See **Note 2**.)
4. Should the biotin-labeling of Cot-1 found to be satisfactory, both the biotin-labeled human Cot-1 DNA and the ligated DNA are then combined and purified together using the QIAGEN QIAquick spin columns as in **step 3** of **Subheading 3.2.1**.

Table 7
Binding Buffer for Cot-1 Subtraction

Reagents	Per sample (µL)	Per 8 samples (µL)
Water	730.5	5844
1 M Tris, pH 8.0	3.75	30
5 M NaCl	15	120
0.5 M EDTA	0.75	6
Total volume	750	6000

5. In order to hybridize Cot-1 to the sample DNA, the dried combined DNA is reconstituted with 6 µL water and 3 µL 20X SSC. This mixture is layered with mineral oil and hybridization is carried out in a thermocycler as follows:
   Step 1: 100°C for 3 min → Step 2: 95°C for 5 min →
   Step 3: 65°C for 16 h → Step 4: 4°C to ∞.
   When thermocycler program reaches Step 3, 1 µL 1% SDS is added to the DNA mixture. This step should be carried out as swiftly as possible.
6. Though the previous step hybridizes biotin-labeled human Cot-1 DNA to the repetitive sequences in the primer-ligated DNA sample, the subtraction step relies upon the high-affinity binding between biotin and streptavidin-magnetic particles to remove those repetitive elements. A 1X binding buffer is prepared. (*See* **Table 7**.)
7. The streptavidin-magnetic particles are thoroughly mixed and rinsed in the 1X binding buffer before being used. Add 100 µL magnetic beads to two 0.6-mL microcentrifuge tubes for each DNA sample. Place the tubes on the magnetic stand and allow the beads to be tightly drawn to the magnet before removing the clear suspension solution. Add 200 µL 1X binding buffer to the beads and re-suspend the beads by mixing. Centrifuge the tubes briefly and replace the tubes onto the magnetic stand once again, allowing for the beads to be tightly drawn by the magnet. Remove the clear binding buffer and cap the tubes tightly until they are ready to be used.
8. Carefully remove the mineral oil from the 10 µL Cot-1 hybridized DNA sample and to it, add 140 µL 1X binding buffer. Mix thoroughly and transfer the sample to the first tube of rinsed streptavidin beads. Allow for the binding of biotin to streptavidin for 20 min with intermittent mixing at room temperature. To remove the hybridized repetitive sequences, place the tube on the magnetic stand where the streptavidin beads would draw the hybridized complexes away. The clear sample suspension is then transferred to the second tube of rinsed streptavidin beads. Repeat the incubation step for another 20 min. The Cot-1 subtracted sample DNA is then purified using the QIAquick PCR Purification Kit, where it is eluted twice with 50 µL water. Dry the purified DNA under vacuum.

Table 8
Reaction Mix for Amplifying Restricted DNA

Reagents	Per sample (µL)	Per 8 samples (µL)
Water	27.4	219.2
DMSO	4	32
10X ThermoPol Buffer	4	32
H-24 (10 µM)	1	8
dNTP (10 mM each)	0.8	6.4
Deep Vent$_R$ ($^{exo-}$) DNA Polymerase (2 U/µL)	0.8	6.4
Total volume	38	304

3.2.4. Methylation-Sensitive Restriction and Amplification

1. Reconstitute the dried DNA with 20 µL water. Take 8 µL DNA and mix it with 1 µL *Bst*UI (10 U/µL) and 1 µL 10X NEB 2 buffer. Layer the mixture with mineral oil and incubate the tube in a 60°C water bath overnight. Save the remaining 12 µL DNA and label the tube as "*Mse*I digested-Cot-1-subtracted (MLC)" for future use. (*See* **Note 3**.)
2. Two microliters (2 µL) of the *Bst*UI digested DNA is PCR amplified by adding 38 µL of the mixture listed in **Table 8**.
   Step 1: 72°C for 5 min → Step 2: 97°C for 1 min →
   Step 3: 72°C for 3 min → Step 4: Go to Step 2 for 14–19 times →
   Step 5: 72°C for 10 min → Step 6: 4°C to ∞. (*See* **Note 4**.)
3. These PCR products are purified with the QIAquick PCR Purification Kit and the purified DNA is eluted twice with 40 µL water. The amplified DNA is dried under vacuum and reconstituted with 50 µL water.

3.3. Array Hybridization

1. Before the amplicons are hybridized onto the CpG island arrays, these sample DNA are labeled first with isotopes. Labeling is carried out with the NEBlot® kit listed in **Table 9**. Sixteen and a half microliters (16.5 µL) of the purified DNA is first denatured at 95°C for 5 min in a 0.6 mL microcentrifuge tube and placed on ice for 2 min.
2. Eight and a half microliters (8.5 µL) of the labeling solution is added to the denatured sample and the reaction is incubated at 37°C for 1 h. While the labeling reaction is underway, prehybridize the CpG island array panel in 7 mL of HEHS at 65°C for at least 40 min. Also, warm another 3 mL of the HEHS solution at 65°C for each sample in a 15-mL polypropylene tube.
3. At the completion of the 1-h labeling reaction, purify the amplicon DNA by eluting it through a G-50 Sephadex column. To determine the labeling efficiency of the amplicon probe, transfer 1 µL of this ^{32}P-labeled probe to a microcentrifuge

Table 9
Reaction Mix for Labeling Amplicon with Radioactive Isotope

Reagents	Per sample (µL)	Per 8 samples (µL)
d(GAT)TP (0.5 mM each)	3	24
10X Labeling buffer	2.5	20
[α]^{32}P-dCTP (3,000 Ci/mmol)	2.5	20
DNA Polymerase I-Klenow fragments	0.5	4
Total volume	8.5	68

tube and read its activity on a scintillation counter. Only probes with a total of at least 2×10^6 cpm should be used for hybridization.

4. Fifty microliters (50 µL) of human Cot-1 DNA (1 mg/mL) is added to the labeled amplicon probe as a suppressor of repetitive sequences (*see* **Note 5**) and the DNAs are denatured at 95°C for 5 min. After chilling it on ice for 2 min, it is added to the 3 mL pre-warmed HEHS. Continue incubating this hybridization mixture at 65°C for another 20 min. This radiolabeled probe mixture is then added to the pre-hybridized membrane of arrays and incubated overnight in a 65°C hybridization oven.

5. Posthybridization washing consists of a 20-min wash at 65°C with 0.1% SDS-0.5X SSC and three 20-min washes at 65°C with 0.1% SDS-0.2% SSC. Membranes are then air-dried and exposed to X-ray films or PhosphorImager cassettes for varying amounts of time. (*See* **Note 6**.)

4. Notes

1. The quality and the quantity of the DNA used in the amplicon generation protocol needs to be well-controlled, and the recommended 0.2–0.5 µg DNA should be followed closely. The test gel for the ligation PCR is revealing in that insufficient starting DNA would result in bright primer-dimer bands with most of the intensity observed at the low molecular-weight region. Too much starting DNA or unsuccessful *Mse*I digestion would result in prominent bands in the high molecular-weight region. When these conditions occur, adjust the DNA concentration accordingly until the ligation test PCR appears as an even smear with intensity around 0.2–2.0 kb.

2. Human Cot-1 DNA subtraction is one of the critical steps in the amplicon generation protocol. As reaction conditions vary somewhat with every set of amplicons generated, it is important to document the efficiency of the entire Cot-1 subtraction step as described in the tracer reaction. Nine microliters (9 µL) of the Nick Translation reaction mixture (**step 2, Subheading 3.2.3.**) is pipetted to a separate 0.6 mL microcentrifuge tube containing 1 µL ^{32}P-dCTP (3,000 Ci/mmol). Incubate this tube together with the tube of reaction master mix at 14°C for the same amount of time. The tracer reaction is passed through a prepared G-50 column after adjusting the volume to 25 µL and the eluate is counted to

determine the efficiency of radiolabeling. Prepared streptavidin particles are incubated twice with this ^{32}P- and biotin-labeled Cot-1 DNA for 20 min with intermittent mixing. After the second incubation period, streptavidin particles are held immobile on the magnetic stand and the clear fraction is removed and counted. It is our experience that when experimental conditions are well-controlled, more than 99.5% of the ^{32}P-labeled, biotin-labeled Cot-1 DNA is removed by the two streptavidin-binding steps.

3. *Bst*UI, a four-base cutter that is sensitive to methylation for restriction (i.e., cuts CGCG, but not ^{m}CGCG; ^{m}C; 5-methylcytosine), is a choice enzyme for the DMH analysis because ~70% of known CpG islands contain its recognition site. Alternatively, other 4-base methylation-sensitive restriction enzymes such as *Hpa*II or *Hha*I are also suitable for DMH.

4. In **step 2** of **Subheading 3.2.4.**, the use of low amplification cycles is important to prevent the over amplification of leftover repetitive sequences by PCR. The recommended 15–20 cycles allow optimal amplification of unique sequences without excessively amplifying the remaining repetitive sequences.

5. Two additional means to control against the nonspecific hybridization of repetitive sequences to the CGI panels are the addition of Cot-1 DNA (50 µg) to ^{32}P-labeled amplicon and the stringent washing temperature at 65°C.

6. It is recommended that several testing steps be performed occasionally to ensure the quality of DNA amplicons as well as the CGI clones selected for the panels. One such step is to hybridize the MLC-only control sample from **step 1, Subheading 3.2.4.** to the CGI panels. At this stage, the amplicons have not yet undergone *Bst*UI digestion, thus they should hybridize to nearly all the clones on the membrane. Another step is to directly hybridize repetitive sequences, such as Cot-1 or ribosomal DNA, to the selected CGI panels and document their hybridization intensities. Having these two pieces of information would greatly help to interpret experimental results in that a lesser credence then may be given to hyper- and hypomethylation of clones that are moderately or strongly positive to these repetitive DNA.

Acknowledgment

This work was supported by grants from the National Cancer Institute CA-69065, CA-84701, and U.S. Army Medical Research Command DAMD17-98-1-8214.

References

1. Benton, W. D. and Davis, P. D. (1977) Screening lambda gt recombinant clones by hybridization to single plaques in situ. *Science* **196**, 180–182.
2. Hara, E. T., Kato, T., Nakada, S., Sekiya, S., and Oda, K. (1991) Subtractive cDNA cloning using oligo(dT)30-latex and PCR: isolation of cDNA clones specific to undifferentiated human embryonal carcinoma cells. *Nucleic Acid Res.* **19**, 7097–7104.

3. LaRosa, G. J. and Gudas, L. J. (1988) An early effect of retinoic acid: cloning of an mRNA (Era-1) exhibiting rapid and protein synthesis-independent induction during teratocarcinoma stem cell differentiation. *Proc. Natl. Acad. Sci. USA* **85,** 329–333.
4. Schena, M., Shalon, D., Davis, R. W., and Brown, P. O. (1995) Quantitative monitoring of gene expression patterns with a complementary DNA microarray. *Science* **270,** 467–470.
5. Ramsay, G. (1998) DNA chips: State-of -the-art. *Nature Biotechnol.* **16,** 40–44.
6. Chen, J. J. W., Wu, R., Yang, P. C., Huang, J. Y., Sher, Y. P., Han, M. H., et al. (1988) Profiling expression patterns and isolating differentially expressed genes by cDNA microarray system with colorimetry detection. *Genomics* **51,** 313–324.
7. Huang, T. H.-M., Perry, M. R., and Laux, D. E. (1999) Methylation profiling of CpG islands in human breast cancer cells. *Hum. Mol. Genet.* **8,** 459–470.
8. Baylin, S. B., Herman, J. G., Graff, J. R., Vertino, P. M., and Issa, J.-P. (1997) Alterations in DNA methylation: a fundamental aspect of neoplasia, in *Advances in Cancer Research*, vol. 72, (Vande Woude, G. F. and Klein, G., eds.), Academic Press, San Diego, CA, pp. 141–196.
9. Zingg, J.-M. and Jones, P. A. (1997) Genetic and epigenetic aspects of DNA methylation on genome expression, evolution, mutation and carcinogenesis. *Carcinogenesis* **18,** 869–882.
10. Cross, S. H., Charlton, J. A., Nan, X., and Bird, A. P. (1994) Purification of CpG islands using a methylated DNA binding column. *Nature Genet.* **6,** 236–244.

9

Methylated CpG Island Amplification for Methylation Analysis and Cloning Differentially Methylated Sequences

Minoru Toyota and Jean-Pierre J. Issa

1. Introduction
1.1. Methylated CpG Island Amplification

CpG islands are clusters of CpG dinucleotides that can be found in the 5′ region of about half of human genes *(1)*. Methylation of cytosine within the 5′ CpG islands is associated with transcriptional inactivation of the involved gene. Aberrant methylation of CpG islands is an important mechanism of gene inactivation in cancer *(2,3)* and other states such as aging and age-related diseases *(4)*. Such methylation often involves numerous CpG islands, and emerging data suggests that methylation profiles may be of some utility in disease detection, prognosis, and risk assessment. However, the measurement of DNA methylation abnormalities at multiple gene loci is currently cumbersome, time-consuming, and not easily amenable to automation. Furthermore, the identification of novel gene sequences hypermethylated in cancer is relatively difficult and inefficient using current technology *(5,6)*. Methylated CpG island Amplification (MCA) was developed to overcome these problems *(7)*.

The principle underlying MCA involves amplification of closely spaced methylated SmaI sites to enrich for methylated CpG islands. The MCA technique is outlined in **Fig. 1**. About 70–80% of CpG islands contain at least two closely spaced (<1kb) SmaI sites (CCCGGG). Only those SmaI sites within these short distances can be amplified using MCA, ensuring representation of the most CpG rich sequences. Briefly, DNA is digested with SmaI, which cuts only unmethylated sites, leaving blunt ends between the C and G. DNA is then digested with the SmaI isoschizomer XmaI, which does cut methylated

From: *Methods in Molecular Biology, vol. 200: DNA Methylation Protocols*
Edited by: K. I. Mills and B. H. Ramsahoye © Humana Press Inc., Totowa, NJ

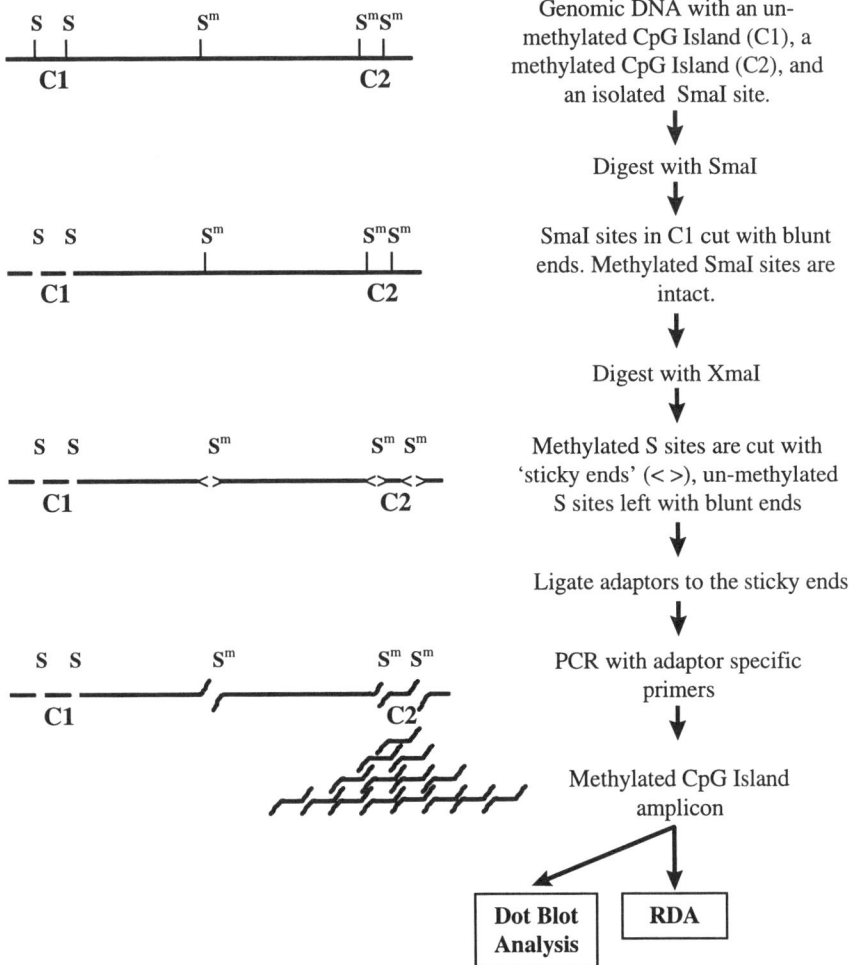

Fig. 1. Outline of the MCA procedure. S, unmethylated SmaI site; S^m Methylated SmaI site.

CCCGGG sites, and which leaves a 4-base overhang. Adaptors are ligated to this overhang, and PCR is performed using primers complementary to these adaptors. The amplified DNA is then spotted on a nylon membrane and can be hybridized with any probe of interest.

To identify CpG islands differentially hypermethylated in cancer (or any appropriate condition); MCA amplicons can be used as templates for subtraction techniques such as representational difference analysis (RDA) *(8)*, using DNA from cancer as tester and DNA from normal tissue as driver. Because this combination is positive selection, contamination of tumor samples with normal cells has little unfavorable effects for the subtraction. In addition to disease-related genes, this technique could potentially be used to identify novel imprinted genes, as well as genes on the X-chromosome.

1.2. Advantages and Disadvantages of MCA

Compared to other techniques used to detect methylation and/or differentially methylated genes, MCA has several advantages: 1) A large number of samples can be analyzed rapidly at multiple loci, 2) many steps of MCA can be readily automatable, 3) MCA allows for an unbiased representation of CpG islands without requiring prior knowledge of their DNA sequence, and 4) novel differentially methylated CpG islands can be amplified and cloned relatively simply, without the need for acrylamide gels or two-dimensional (2-D) gel electrophoresis. However, MCA also has a few potential disadvantages in that 1) it requires relatively good-quality DNA, precluding the use of paraffin-embedded samples, 2) it examines only a limited number of CpG sites within CpG islands, 3) CpG islands that do not contain two closely spaced (1 kb) SmaI sites cannot be analyzed, 4) false-positives can result from incomplete digestion using the methylation-sensitive restriction-enzyme SmaI.

2. Materials
2.1. MCA

1. Restriction enzymes SmaI, XmaI.
2. T4 DNA ligase.
3. Taq DNA polymerase.
4. 10X polymerase chain reaction (PCR) reaction buffer: 670 mM Tris-HCl, pH 8.8, 40 mM MgCl$_2$, 160 mM NH$_4$(SO$_4$)$_2$, 100 mM β-mercaptoethanol, 1 mg/mL bovine serum albumin (BSA).
5. Tris-EDTA (TE), pH 8.0.
6. DNA precipitation reagents: Phenol/Chloroform, pH 8.0–9.0, 3M NaOAc (for general precipitation); 5 M NH$_4$OAc (for precipitation and quantitation when dNTPs are present); 100% ETOH.
7. Agarose gel electrophoresis reagents.
8. Filter hybridization reagents: 96 pin replicator system (Nunc), Nylon membranes, DNA hybridization solution (e.g., BLOTTO), Random-primed DNA labeling kit, Wash solutions: Wash 1 2X SSC, 0.1% SDS; Wash 2 0.1X SSC, 0.1% SDS.

2.2. RDA and Cloning PCR Products

1. 3X EE buffer: 30 mM EPPS (Sigma), pH 8.0, 3 mM ethylenediaminetetraacetic acid (EDTA), pH 8.0.
2. 5 M NaCl.
3. cDNA spun column (Amersham).
4. Mung bean nuclease (NEB).
5. pBluescript (Stratagene).

2.3. Oligonucleotides

RXMA primers
    RXMA24 : 5′-AGCACTCTCCAGCCTCTCACCGAC-3′
    RXMA12 : 5′-CCGGGTCGGTGA-3′
    JXMA24 : 5′-ACCGACGTCGACTATCCATGAACC-3′
    JXMA12 : 5′-CCGGGGTTCATG-3′
    NXMA24 : 5′-AGGCAACTGTGCTATCCGAGTGAC-3′
    NXMA12 : 5′-CCGGGTCACTCG-3′

RMCA primers
    RMCA24 : 5′-CCACCGCCATCCGAGCCTTTCTGC-3′
    RMCA12 : 5′-CCGGGCAGAAAG-3′
    JMCA24 : 5′-GTGAGGGTCGGATCTGGCTGGCTC-3′
    JMCA12 : 5′-CCGGGAGCCAGC-3′
    NMCA24 : 5′-GTTAGCGGACACAGGGCGGGTCAC-3′
    NMCA12 : 5′-CCGGGTGACCCG-3′

3. Methods

3.1. Preparation of MCA Amplicons

3.1.1. Digestion of Genomic DNA

1. Digest 5 µg of genomic DNA using 100 units of SmaI over night.
2. Add 20 units of XmaI and incubate at 37°C for 6 h.
3. Add one volume PC9, vortex, spin, and extract the supernatant.
4. Precipitate the DNA: Add 1/10th volume 3 M NaOAc and 2 volumes 100% ETOH. Store at –70°C for 1 h and centrifuge 30 min at >10,000g. Pour the ETOH out, air-dry the pellets.
5. Resuspend in 10–20 µL TE and determine DNA concentration using a spectrophotometer.

3.1.2. Ligation of Adapter

1. Prepare an adaptor mixture by diluting the primers to 100 µM and combining 50 µL of RXMA24 with 50 µL RXMA12 (or RMCA24 and RMCA12).
2. Incubate at 65°C for 2 min and cool to room temperature for 30–60 min. This mixture can be stored at –20°C for up to 6 mo.
3. Mix the following: 500 ng of Digested DNA, 10 µL of adaptor mixture, 400 Units of T4 DNA ligase, 3 µL of 10X ligase buffer and water to a total volume of 30 µL.
4. Incubate at 16°C for 3–16 h.

3.1.3. PCR Amplification

1. Prepare tubes containing 10 µL of 10X PCR buffer, 100 pmol of RXMA24 (or RMCA24) primers, 15 Units of Taq DNA polymerase, 1.2 µL dNTP mix (25 mM), 0 µL (RXMA) or 5 µL (RMCA) DMSO, H$_2$O to a total volume of 97 µL.

2. Add 3 µL of ligation mixture. Cover with mineral oil.
3. To fill the 3′-recessed ends of the ligated fragments, incubate at 72°C for 5 min.
4. Perform 25 cycles of PCR (95°C for 1 min and 72°C [for RXMA24] or 77°C [for RMCA24] for 3 min), with a final extension time of 10 min.
5. After the reaction, electrophorese 10 µL of the PCR products in a 1.5% agarose gel to check the quality of the amplification. You should see a relatively strong smear, ranging from 300 bp to 2 kb.

3.2. Detection of Aberrant Methylation by Dot-Blot Analysis

3.2.1. Preparation of Filters

1. Transfer the PCR products to a new tube and add 2/3 vol of 5 M NH$_4$OAc and 350 µL of 100% ethanol.
2. Chill at –70°C for 1 h and precipitate DNA by centrifugation. Resuspend DNA in 10–15 µL TE buffer and quantitate in a spectrophotometer.
3. Dilute India Black Ink by adding 20 µL to 10 mL H$_2$O. Add 20 µL of this diluted solution to 10 mL 20X SSC.
4. Dilute 1 µg of MCA amplicon in TE (total volume 4 µL). Add 2 µL of the 20X SSC/India Ink solution.
5. Blot (in duplicate) onto nylon membranes. The easiest way to do this is to transfer the DNA mix to a 96-well plate and use a 96-pin replicator system (Nunc). Dry the membrane at room temperature (RT) for 30 min.
6. Place the filters in 0.5 M NaOH/1.5 M NaCl solution for 5 min.
7. Place the filters in 0.5 M Tris-HCl, pH 8.0/1.5 M NaCl. Neutralize for 5 min.
8. Transfer the filters to 3 M SSC and rinse for 5 min.
9. Dry the filters at RT for 1 h.
10. Cross link the DNA to the filters using a UV cross-linker (or bake at 80°C for 30 min).

3.2.2. Hybridization

1. Prehybridize filters in a DNA prehybridization solution such as Blotto at 65°C for 3 h.
2. Label 20 ng of the probe using random priming and ^{32}P dCTP. Boil the probe, cool on ice, and add to the hybridization solution.
3. Hybridize filters for 12–16 h.
4. Wash with 2X SSC, 0.1% SDS at 65°C for 10 min twice, and 0.1X SSC, 0.1% SDS at 65°C for 20 min.
5. Expose the filters to a phosphor screen (or use conventional film autoradiography).
6. Develop after 1–3 d exposure. See examples of results in **Fig. 2**.

3.3. MCA Coupled with RDA

3.3.1. Outline

For detection of differentially methylated sequences, it is necessary to generate MCA amplicons from the tester samples (e.g., a cancer sample) and

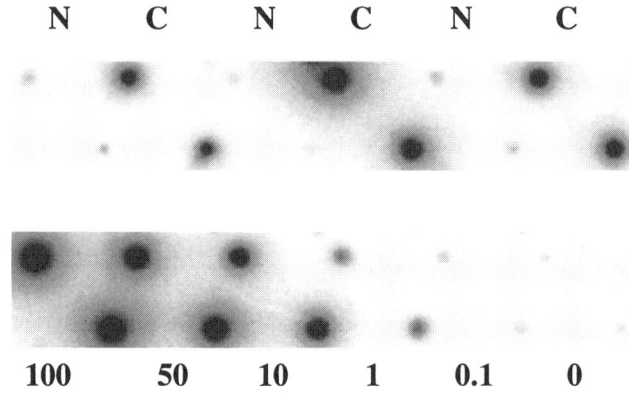

Fig. 2. Example of dot-blot analysis of MCA products. MCA was performed as described in the text on three pairs of colon cancer (C) and adjacent normal mucosa (N) (top panel). For controls, we used MCA products from the colon-cancer cell line Caco2 and a normal colon sample mixed in varying proportions (bottom panel). The numbers below the bottom panel represent the percentage of Caco2 DNA in the samples (prior to MCA). Thus, the lane labeled 100 has only Caco2 DNA, the lane labeled 0 has only normal colon DNA, and the lane labeled 50 contains an equal mixture of Caco2 and normal colon. The blots were probed with MINT31, a DNA fragment obtained by MCA/RDA using Caco2 as the tester and normal colon as driver. Note that MCA appears to detect MINT31 methylation semi-quantitatively (bottom panel), and that MINT31 is methylated in all three cancer samples but in less than 1% of the normal adjacent colon.

relatively large quantities of MCA amplicons from the driver samples (e.g., DNA from normal tissues) with the adaptors removed. The tester's adaptors will then be changed and the DNA hybridized with driver DNA, followed by PCR amplification using the second set of adaptors. The subtraction is then repeated once, and the resulting amplicons are further cloned and characterized.

3.3.2. Removal of Adaptors from the Driver Amplicon

1. Perform MCA on multiple aliquots of driver DNA. We typically run 10 reactions in parallel. Verify and pool the aliquots. Quantitate in a spectrophotometer.
2. Digest the driver MCA amplicons using 2 Units/µg SmaI to remove the RMCA/RXMA adaptor. Inactivate SmaI by phenol/chloroform extraction.
3. Remove the adaptors using a cDNA Spun column (Amersham).
4. Electrophorese an aliquot of DNA before and after column filtration to check for complete elimination of adaptors.

5. Add 1/30th volume of 3 M Sodium acetate, two volumes of ethanol, chill at −70°C, precipitate by centrifugation, resuspend in 200–400 µL TE, and quantitate by spectrophotometry.
6. 80 µg of processed driver DNA is required per tester sample (per condition).

3.3.3. Change of Adaptors on the Tester Amplicon

1. Digest 5 µg of tester MCA amplicons using 20 Units XmaI to remove the RMCA/RXMA adaptor. Inactivate XmaI by phenol/chloroform extraction.
2. Remove the adaptors using a cDNA Spun column (Amersham).
3. Electrophorese an aliquot of DNA before and after column filtration to check for complete elimination of adaptors.
4. Add 1/30th volume of 3 M Sodium acetate, two volumes of ethanol, chill at −70°C, precipitate by centrifugation, resuspend in 10–20 µL TE, and quantitate by spectrophotometry.
5. Prepare JMCA and/or JXMA adaptors as described in **Subheading 3.1.2.** Ligate the adaptors to 0.5 µg of the above DNA as described in **Subheading 3.1.2.**

3.3.4. Competitive Hybridization

1. Add 70 µL of TE to the ligation mix and purify DNA by phenol/chloroform extraction. Mix all of this DNA with 40 µg of MCA amplicons from driver DNA.
2. Add 1/10th volume of 3 M NaOAc and 2 vol of ethanol, chill at −70°C for 30 min, and centrifuge for 30 min.
3. Dissolve DNA in 4 µL of 3X EE (30 mM EPPS, 3 mM EDTA) solution and transfer to a 0.5-mL microcentrifuge tube.
4. Denature DNA at 96°C for 10 min, quick spin, add 1 µL of 5 M NaCl, cover with mineral oil, and incubate at 67°C for 20 h (this is best done in a thermocycler).

3.3.5. Selective Amplification

1. Heat 100 µL of 1 M NaCl at 67°C for 5 min, and add 45 µL to the competitive hybridization solution.
2. Prepare PCR mixture as follows: 10 µL of 10X PCR buffer, 5 µL (1/10th) of the hybridization mix; 1.2 µL of 25 mM dNTP mix, 100 pmol JXMA24 (or JMCA24) primers, 15 Units of Taq DNA polymerase, 0 µL (RXMA) or 5 µL (RMCA) dimethylsulfoxide (DMSO), H$_2$O to a total volume of 100 µL. Cover with mineral oil.
3. Fill the ends at 72°C for 5 min. PCR amplification is then done at 10 cycles of 95°C for 1 min and 72°C (JXMA24) or 77°C (JMCA24) for 3 min. Final extension is at 72°C for 10 min.
4. Transfer the PCR products to a clean microcentrifuge tube.

5. To digest the single-stranded amplified MCA products, add 10 µL of 10X Mung bean nuclease buffer and 100 units of Mung bean nuclease, and incubate at 30°C for 30 min.
6. Purify DNA with phenol/chloroform extraction, add 2/3 vol of NH_4OAc, and two volumes of ethanol, chill at –70°C for 30 min and precipitate by centrifugation.
7. Resuspend DNA in 50 µL water. Add 10 µL of 10X PCR buffer, 1.2 µL of 25 m*M* dNTP mix, 100 pmol JXMA24 (or JMCA24) primers, 15 Units of Taq DNA polymerase, 0 µL (RXMA) or 5 µL (RMCA) DMSO, H_2O to a total volume of 100 µL. Cover with mineral oil.
8. PCR amplification is then done at 20 cycles of 95°C for 1 min and 72°C (JXMA24) or 77°C (JMCA24) for 3 min. Final extension is at 72°C for 10 min.
9. After the reaction, electrophorese 10 µL of the PCR products in a 1.5% agarose gel to check the quality of the amplification. You should see a DNA smear, ranging from 200 bp to 1 kb.

3.3.6. Second-Round Subtraction

1. Remove JMCA/JXMA adaptors and ligate NMCA/NXMA adaptors as described in **Subheading 3.3.3.**
2. For the second round of subtraction 100 ng of tester and 40 µg of driver are used. Hybridize and proceed with PCR as outlined in **Subheadings 3.3.4.** and **3.3.5.**
3. Clone the MCA products obtained after two rounds of subtraction and amplification: DNA is digested with XmaI, purified by phenol/chloroform extraction and ethanol precipitation.
4. Resuspend the DNA in water and ligate to pBluescript (or your favorite cloning vector) digested with XmaI and treated with calf intestinal phosphatase. Transform appropriate strains of bacteria, grow overnight and screen the colonies for the presence of inserts.

3.3.7. Identification of Differentially Methylated Sequences

1. At this point, many of the inserts are Alu repetitive elements, and some are false-positives.
2. To identify nonrepeated true positives, we first amplify the inserts by PCR and screen them by dot-blot hybridization using an Alu probe.
3. All non-Alu hybridizing inserts are then used as probes on filters containing MCA amplicons from tester and driver.
4. All inserts that detect a strong dot in the tester and a weak (or absent) dot in the driver are sequenced and further characterized by Southern blotting and other techniques.

4. Notes

1. CpG islands vary in their CpG density such that different PCR primers and conditions may be required for effective MCA amplification. For this reason, we

have developed two different sets of primers (RMCA and RXMA) that differ in CG content and represent slightly different subsets of CpG islands. On average, the RMCA primers amplify smaller and more CG rich fragments than RXMA. Some probes work well using either condition, but others work better (or exclusively) using one of the sets of primers. For example, the P16 CpG island amplifies better with RMCA. Thus, to adapt this procedure to a known gene, one has to try both RMCA and RXMA, probe with a DNA fragment that is contained between two SmaI sites in the gene of interest, and determine the optimal condition. In some cases, it may be required to change the PCR conditions and/or primers. Similarly, for cloning differentially methylated CpG islands, we have used both RMCA and RXMA amplicons, and have obtained a different spectrum of sequences.

2. It is highly recommended to add positive and negative controls for MCA on every filter. For semi-quantitative detection of methylation by MCA, we also recommend adding a mixture of the positive and negative controls on every filter. For example, we usually use a mixture of a colon-cancer cell line (Caco2) and normal colon DNA, where the proportion of the cancer DNA (which is methylated at the locus of interest) is 100, 50, 10, 1, 0.1, and 0%. The intensity of hybridization for the unknown samples can then be compared to this dilution curve (*see* **Fig. 2**).

3. In our experience, 70–90% of the amplified MCA products represents Alu sequences. After two rounds of subtraction, 50–70% of the subtracted fragments also contains Alu sequences. This may be due to the fact that some Alu's are differentially methylated in cancer. When excluding Alu sequences, 70–80% of the fragments were true-positives, i.e., differentially methylated. The rest were sequences methylated in both testers and drivers. Using MCA/RDA, we have not recovered sequences that are unmethylated in both tester and driver.

4. Taq Gold (PE) does not work well in these assays. Use regular Taq instead.

References

1. Bird, A. P. (1986) CpG-rich islands and the function of DNA methylation. *Nature* **321,** 209–213.
2. Baylin, S. B., Herman, J. G., Graff, J. R., Vertino, P. M., and Issa, J. P. (1998) Alterations in DNA methylation: a fundamental aspect of neoplasia. *Adv. Cancer Res.* **72,** 141–196.
3. Schmutte, C. and Jones, P. A. (1998) Involvement of DNA methylation in human carcinogenesis. *Biol. Chem.* **379,** 377–388.
4. Issa, J. P., Ottaviano, Y. L., Celano, P., Hamilton, S. R., Davidson, N. E., and Baylin, S. B. (1994) Methylation of the oestrogen receptor CpG island links ageing and neoplasia in human colon. *Nat. Genet.* **7,** 536–540.
5. Akama, T. O., Okazaki, Y., Ito, M., Okuizumi, H., Konno, H., Muramatsu, M., et al. (1997) Restriction landmark genomic scanning (RLGS-M)-based genome-wide scanning of mouse liver tumors for alterations in DNA methylation status. *Cancer Res.* **57,** 3294–3299.

6. Gonzalgo, M. L., Liang, G., Spruck, C. H. 3rd., Zingg, J. M., Rideout, W. M. 3rd., and Jones, P. A. (1997) Identification and characterization of differentially methylated regions of genomic DNA by methylation-sensitive arbitrarily primed PCR. *Cancer Res.* **57,** 594–599.
7. Toyota, M., Ho, C., Ahuja, N., Jair, K-W., Ohe-Toyota, M., Herman, J. G., et al. (1998) Epigenetic instability in colorectal cancer. Submitted.
8. Lisitsyn, N., Lisitsyn, N., and Wigler, M. (1993) Cloning the differences between two complex genomes. *Science* **259,** 946–951.

10

Isolation of CpG Islands Using a Methyl-CpG Binding Column

Sally H. Cross

1. Introduction

Vertebrate genomes are globally heavily methylated at the sequence CpG, with the exception of short patches of GC-rich DNA of between 1–2 kb in size that are free of methylation, and these are known as CpG islands (*see* **refs.** *1* and *2* for reviews). In addition to distinctive DNA characteristics, CpG islands also have an open chromatin structure in that they are hyperacetylated, lack histone H1, and have a nucleosome-free region *(3)*. The major reason for interest in CpG islands is that they co-localize with the 5′ end of genes. Both promoter sequences and the 5′ parts of transcription units are found within CpG islands. It has been estimated that 45,000 (56%) of human genes and 37,000 (47%) of mouse genes are associated with a CpG island *(4)* and these include all genes that are ubiquitously expressed as well as many genes with a tissue-restricted pattern of expression *(5,6)*.

Generally CpG islands remain methylation-free in all tissues, including the germ-line regardless of the activity of their associated gene. There are three major exceptions to this. These are CpG islands on the inactive X chromosome *(7)*, CpG islands associated with nonessential genes in tissue-culture cell lines *(8)* and CpG islands associated with some imprinted genes *(9)*. Epigenetic gene silencing associated with aberrant methylation of CpG islands is found in both cancer and ageing *(10,11; see* **ref.** *12* for a review). Why CpG islands are protected from methylation is not clear, but deletion of functional Sp1 binding sites from both the mouse and hamster *Aprt* genes result in CpG island methylation suggesting that such functional transcription-factor binding sites are involved *(13,14)*. This notion is supported by studies examining two of the rare CpG islands not found at the 5′ end of a gene *(15,16)*. In both cases, transcripts arising from the CpG-island region have been identified *(17,18)*. Replication of CpG islands during early S phase has also been suggested to be involved in the protection of CpG island from methylation as it has been found that replication origins are often found at CpG islands *(19)*.

The unusual base composition and methylation-free status of CpG islands enables their detection by restriction enzymes whose sites are rare and, if present, usually blocked by methylation in the rest of the genome *(20)*, and this has proved useful for both mapping and identification of CpG island genes. Here a method is described by which largely intact CpG islands can be isolated from both whole genomes and cloned DNAs by exploiting the differential affinity of DNA fragments containing different numbers of methyl-CpGs for a methyl-CpG binding domain (MBD) column *(21)*. These columns consist of the MBD of the protein MeCP2 *(22,23)* coupled to a resin. MeCP2 is one of a family of proteins that binds symmetrically methylated CpGs in any sequence context and is involved in mediating methylation-dependent repression *(22,24–26)*. DNA encoding the MBD was cloned into a bacterial expression vector to give plasmid pET6HMBD, which, when expressed, yields a recombinant protein, HMBD, consisting of the MBD preceded by a tract of six histidines *(21)*. This histidine tag at the N-terminal end enables the HMBD protein to be coupled to a nickel-agarose resin, which can be packed into a column. DNA fragments containing many methylated CpGs bind strongly and unmethylated DNA fragments bind weakly to MBD columns *(21)*. On average, within CpG islands CpGs occur at a frequency of 1/10 bp and are unmethylated, whereas outside CpG islands CpGs are found at a frequency of 1/100 bp and are usually methylated. CpG islands are 1–2 kb in size and therefore contain between 100–200 CpGs. When unmethylated, as is usually the case in the genome, they show little affinity for binding to MBD columns. However, when methylated, they bind strongly and can be purified away from other genomic fragments that contain few methylated CpGs and therefore bind weakly.

Using MBD columns CpG island libraries have been made for several species *(21,27–29)*. Use of these libraries as probes has shown that, in each case, CpG islands are nonrandomly distributed such they are concentrated in early-replicating, highly acetylated genomic domains *(27–30* and *see* also **ref. 31***)*. The representation of each CpG island in these libraries is unaffected by the expression pattern of its associated gene. Therefore CpG island libraries contain an unbiased collection of fragments of the 5′ ends of many genes, which, because they overlap the 5′ end of the transcription unit, can be used to identify the associated full-length cDNAs either by screening cDNA libraries or searching sequence databases. When used as probes, CpG islands are usually found to be single-copy which facilitates their use to both isolate cDNAs and to map genes. As they contain promoter sequences and therefore transcription-factor binding sites, they can be screened for genes controlled by a particular transcription factor *(32)*. MBD columns have also been used to isolate CpG islands from large genomic clones *(33)* and sorted human chromosomes *(34)*. Analysis of the human genomic fraction that binds tightly to the MBD column revealed that the nontranscribed part of the ribosomal repeat was methylated *(35)*. Finally methylation of CpG islands appears to be one route by which genes are epigenetically silenced in cancer (reviewed in **ref. *12***). Such methylated CpG islands have been identified both by screening the human CpG island library *(36)* or by directly isolating methylated CpG islands using the MBD column *(37)*. The different ways MBD columns can be used are discussed in **Subheading 3.5.**

The general protocol can be split into the following steps:

1. Production of HMBD and coupling to nickel-agarose to form the MBD column.
2. Calibration of the MBD column using plasmid DNAs containing known numbers of methyl-CpGs.
3. Restriction digestion of the DNA of interest such that CpG islands are left largely intact and other DNA is reduced to small fragments.
4. Fragments containing clusters of methylated CpGs are removed by passing the DNA over the MBD column and collecting the flow-through or "stripped" DNA. These include GC-rich, non-CpG island sequences such as repeats. In the preparation of CpG-island libraries, this step is important for eliminating such fragments, which would otherwise contaminate the library.
5. The stripped DNA is then methylated at all CpGs.
6. The methylated stripped DNA is then passed over the column again. CpG island fragments now bind strongly. Elution at high salt yields a DNA fraction highly enriched for largely intact CpG islands.

Depending on the particular application, some of the aforementioned steps may be omitted. For example, as methylation is erased on cloning, when isolating CpG islands from cloned DNA the stripping **step 4** is omitted *(33)*. In some cases, it is the DNA present in the fraction that binds tightly in **step 3** which is of interest *(35)*, for example, when isolating aberrantly methylated or imprinted CpG islands *(37)*.

2. Materials
2.1. Preparation of the MBD Column

1. LB broth: 1% bacto tryptone, 0.5% bacto yeast extract and 1% NaCl (all w/v).
2. LB agar: As LB broth with the addition of 12 g/L Bacto agar.
3. 100 mM isopropyl β-D thiogalactopyranoside (IPTG) in water, filter-sterilized. Store at −20°C.

4. 2X SMASH buffer: 125 mM Tris-HCl, pH 6.8, 20% glycerol, 4% sodium dodecyl sulfate (SDS), 1 mg/mL bromophenol blue, 286 mM β-mercaptoethanol. Divide into aliquots, keep the one in use at room temperature, and store the others at –20°C until required.
5. 100 mM phenylmethylsufonyl fluoride (PMSF) in isopropanol. Store at 4°C. Add to buffers A, B, C, D, and E to a final concentration of 0.5 mM just before use.
6. Stock solutions of the following protease inhibitors: leupeptin, antipain, chymostatin, pepstatin A, and protinin prepared and stored as recommended by the manufacturer. Add to buffers A, B, C, D, and E to a final concentration of 5 µg/mL just before use.
7. 20% Triton X-100.
8. Buffer A: 5 M urea, 50 mM NaCl, 20 mM HEPES, pH 7.9, 1 mM ethylenedimainetetraacetic acid (EDTA), pH 8.0, 10% glycerol.
9. Buffer B: 5 M urea, 50 mM NaCl, 20 mM HEPES, pH 7.9, 10% glycerol, 0.1% Triton X-100, 10 mM β-mercaptoethanol.
10. Buffer C: 2 M urea, 1 M NaCl, 20 mM HEPES, pH 7.9, 10% glycerol, 0.1% Triton X-100, 10 mM β-mercaptoethanol.
11. Buffer D: 50 mM NaCl, 20 mM HEPES, pH 7.9, 10% glycerol, 0.1% Triton X-100, 10 mM β-mercaptoethanol.
12. Buffer E: 50 mM NaCl, 20 mM HEPES, pH 7.9, 10% glycerol, 0.1% Triton X-100, 10 mM β-mercaptoethanol, 8 mM immidazole.
13. 1 M immidazole in water, filter-sterilized. Store at room temperature.

2.2. Basic Protocol for Running an MBD Column

1. MBD buffer: 20 mM HEPES, pH 7.9, 10% glycerol, 0.1% Triton X-100.
2. MBD buffer/**x** M NaCl: 20 mM HEPES, pH 7.9, **x** M NaCl, 10% glycerol, 0.1% Triton X-100. The **x** indicates a variable value as specified in each protocol.
3. 5 M NaCl.
4. 100 mM PMSF prepared and stored as in **Subheading 2.1.** Add to MBD buffers to a final concentration of 0.5 mM just before use.

2.3. Calibrating the MBD Column

The reagents required for these protocols are generally available in molecular biology laboratories and an extensive list will not be included here. Specifically, reagents required for DNA isolation, purification, restriction-enzyme treatment, and methylation will be needed. The reagents and the techniques are described in *(38)*.

3. Methods

In **Subheading 3.1.**, the preparation of an MBD column is described. **Subheadings 3.2.** and **3.3.** contain the basic protocol for running an MBD column and how to calibrate it. In **Subheading 3.4.**, the preparation of DNAs to be fractionated is described and in **Subheading 3.5.** the various different ways in which an MBD column can be used are discussed.

3.1. Preparation of the MBD Column

To prepare an MBD column the recombinant protein HMBD is expressed in the *Escherichia coli* strain BL21 (DE3) pLysS, partially purified, coupled to nickel-agarose resin and packed into a column. The T7 RNA polymerase-expression system is used to produce HMBD protein *(39)*. This protocol should produce sufficient HMBD protein to make a 1 mL column, and may be adjusted as required.

All the steps after **step 6** of **Subheading 3.1.1.** are done on ice or in a cold room using ice-cold solutions.

3.1.1. Preparation of HMBD protein

1. Streak BL21 (DE3) pLysS (pET6HMBD) from a –80°C stock onto a LB agar plate containing ampicillin (50 µg/mL) and chloramphenicol (30 µg/mL) and grow overnight at 37°C to obtain single colonies.
2. Innoculate 100 mL LB broth containing ampicillin (50 µg/mL) and chloramphenicol (30 µg/mL) with a single colony. At 37°C shake at about 300 rpm overnight in a 500-mL flask.
3. Inoculate 1.5 L of LB broth containing ampicillin (50 µg/mL) and chloramphenicol (30 µg/mL) with 45 mL of the overnight culture. Measure the OD_{600} (optical density at 600 nm). This should be approx 0.1. If not, adjust accordingly. Split the culture between two 2-L flasks and shake vigorously for 2–3 h at 37°C until the OD_{600} has reached between 0.3 and 0.5. Remove a 500 µL aliquot (sample 1).
4. To each flask, add IPTG to a final concentration of 0.4 m*M*. Grow the cultures for 3 h at 37°C with vigorous shaking. Remove another 500 µL aliquot (sample 2).
5. Centrifuge samples 1 and 2 at 14K (full speed) in a microfuge for 5 min at room temperature. Resuspend the pellets in 100 µL sterile, distilled water plus 100 µL 2X SMASH buffer and store at –20°C until required for the analysis gel.
6. Centrifuge the rest of the cells at 2000*g* for 20 min at 4°C in two 1 L centrifuge bottles.
7. Discard the supernatants and resuspend each pellet in 12.5 mL buffer A. Transfer to a 50 mL tube, add Triton X-100 to 0.1% and mix by gentle swirling. The solution will become viscous as the cells begin to lyse on addition of the Triton X-100.
8. Disrupt the cells and shear the DNA by sonication. The extract will lose its viscosity and may darken in color. Remove a 100 µL aliquot, add 100 µL of 2X SMASH buffer, mix, and store at –20°C (sample 3).
9. Centrifuge the disrupted cells at 31,000*g* for 30 min at 4°C. Pour the supernatant into a 50 mL tube. Remove a 100 mL aliquot, add 100 µL of 2X SMASH buffer, mix, and store at –20°C (sample 4). Store the remaining supernatant (approx 25 mL) at –80°C until required, otherwise go on to **step 3** in **Subheading 3.1.2.**

3.1.2. Partial Purification of the HMBD Protein

To do this, the crude protein extract prepared in **Subheading 3.1.1.** is passed over a cation exchange resin to which most of the contaminating bacterial proteins bind weakly but the basic HMBD protein (predicted pI 9.75) binds tightly.

1. If the protein extract has been stored at –80°C, thaw in cold water or on ice. Add protease inhibitors (**Subheading 2.1., items 5** and **6**) and mix by swirling.
2. To remove insoluble material centrifuge at 31,000g for 30 min at 4°C. Pour the supernatant into a 50 mL tube and discard the pellet.
3. Prepare 12 mL of Fractogel EMD SO3^e-650(M) (Merck) resin as recommended by the manufacturer and pipet 5 mL into each of two plastic disposable chromatography columns, such as Econo-Pac columns (Bio-Rad 732-1010). Attach a syringe needle to each column. This increases the flow rate. Two 5 mL columns are used rather than one 10 mL column to reduce the time taken by this protocol. To equilibrate the columns, wash each with 25 mL buffer B, followed by 25 mL buffer C, and finally with 25 mL buffer B.
4. Arrange the two columns so that they can drip into the same tube. Simultaneously, load half of the supernatant on one column and the other half on the other column. Collect the flow-through (FT) in a single 50 mL tube and keep on ice.
5. Next, elute the bound protein by washing the columns simultaneously in 12 elution steps. For each wash step, collect the eluates from both columns into a single 15 mL tube. For washes 1–4, use 5 mL of buffer B/column; for washes 5–8, use 5 mL of buffer B + C (27.5 mL of buffer B + 12.5 mL of buffer C)/column; and for washes 9–12, use 5 mL of buffer C/column. Keep fractions 1–12 on ice.
6. To ascertain which fractions contain the HMBD protein, remove 10 µL aliquots from each fraction and the FT. Add 10 µL 2X SMASH buffer to each. Heat these samples and samples 1–4 (put aside in **Subheading 3.1.1.**) at 90°C for 90 s. Separate 20 µL of each on a 15% sodium dodecyl sulfate-polyacrylamide gel electrophoresis (SDS-PAGE) gel, along with molecular-weight (MW) standards (for example Protein marker, Broad Range [2–212 kD], New England Biolabs 7701S) and stain with Coomassie Brilliant Blue R-250 using standard techniques *(38)*. The MW of the HMBD is 11.4 kD and should be present in samples 2–4 and fractions 9–12. Pool all fractions enriched for the HMBD protein. The partially purified extract can be stored at –80°C until required, otherwise go to **Subheading 3.1.2.**

3.1.3. Coupling the HMBD Protein to Nickel-Agarose Resin

1. If the protein extract has been stored at –80°C thaw on ice or in cold water. Add protease inhibitors (**Subheading 2.1., items 5** and **6**) and mix by swirling. Remove a 10 µL aliquot and use it to measure the protein concentration by the Bradford assay *(40)* using, for example, the Protein Assay kit (Bio-Rad

500-0002). Typically, the total amount of protein will be about 20–50 mg (approx 1 mg/mL). Remove 50 µL of the protein extract, add 50 µL of 2X SMASH buffer, mix, and keep on ice (sample 5).
2. Pipet 1 mL of nickel-agarose resin (for example Ni-NTA Superflow, Qiagen 30410) into a 5 mL disposable plastic chromatography column (for example Poly-Prep chromatography column Bio-Rad 731-1550). Wash with 4 mL of buffer D to equilibrate.
3. Load the protein extract onto the column and collect the flow-through (FT) in a 50 mL tube.
4. Wash the column with 4 mL of buffer D, followed by 4 mL of buffer E and finally with 4 mL of buffer D, collecting twelve 1 mL fractions.
5. To ascertain if the coupling of the HMBD protein to the nickel-agarose resin has been successful remove 10 µL aliquots from the FT and each fraction. Add 10 µL 2X SMASH buffer to each. Heat these samples and sample 5 at 90°C for 90 s. Separate 20 µL of each on a 15% SDS-PAGE gel, along with molecular-weight standards (for example Protein marker, Broad Range [2-212 kD], New England Biolabs 7701S) and stain with Coomassie Brilliant Blue R-250 using standard techniques *(38)*. If the coupling reaction has been successful, the HMBD protein should be visible in sample 5 but absent or present in trace amounts in the FT and wash fractions.
6. Estimate the amount of HMBD protein coupled to the nickel agarose. Pool the FT and the 12 eluted fractions in a 50-mL tube and measure the protein concentration as in **step 1**. Subtract the amount of protein eluted from the amount of protein loaded to find the amount of HMBD coupled to the resin.
7. Pack the coupled resin into a column.

3.2. Basic Protocol for Running an MBD Column

When fractionating differently methylated DNAs using an MBD column, the same basic procedure is followed and this is outlined here. DNAs are eluted from MBD columns by increasing the NaCl concentration in the wash buffer. Generally, a 1 mL column is suitable for most applications. MBD columns should be run in a cold room using ice-cold solutions. Do not allow the MBD column to dry out.

1. Prepare MBD buffer and MBD buffer/1 M NaCl. Mix these together to make MBD buffers containing the required NaCl concentrations.
2. Equilibrate the MBD column by washing it with 5 column volumes of MBD buffer/ 0.1 M NaCl, followed by 5 column volumes of MBD buffer/1 M NaCl, followed by 5 column volumes of MBD buffer/0.1 M NaCl (*see* **Note 9**).
3. Load the DNA (in MBD buffer/0.1 M NaCl). Wash the column with 5 mL of MBD buffer/0.1 M NaCl (*see* **Note 10**).
4. To elute bound DNAs increase the NaCl concentration present in the wash buffer as either a linear gradient or in steps up to a maximum of 1 M NaCl. This is done

by mixing MBD buffer and MBD buffer/1 M NaCl in the correct proportions (*see* **Note 10**).
5. During **steps 2** and **3**, collect fractions of the size required in the procedure being used. The usual size of the fractions collected is 1 or 2 mL, although in some cases larger volumes are collected.
6. Wash the MBD column with 5 column volumes of MBD buffer/1 M NaCl followed by 5 column volumes of MBD buffer/0.1 M NaCl after use and store at 4°C or in a cold room.

3.3. Calibrating the MBD Column

The amount of HMBD coupled on a MBD column determines the NaCl concentration at which DNAs methylated to different degrees elute. As this varies from column to column each MBD column should be calibrated by determining the elution profile of artificially methylated plasmid DNAs that contain different numbers of methyl-CpGs. To do this a cloning vector such as pUC19, which contains 173 CpGs (accession number M77789), could be used, but any plasmid with a known sequence, and therefore a known number of CpGs, is suitable. Typically, heavily methylated DNA fragments (those containing greater than 100 methyl-CpGs) elute between 0.7 and 0.9 M NaCl. Unmethylated DNA generally elutes at 0.5–0.6 M NaCl (but *see* **Note 10**) and DNAs containing intermediate numbers of methyl-CpGs (30–40) elute at 0.1–0.2 M less than heavily methylated fragments *(21)*.

3.3.1. Preparation of the Differentially Methylated Plasmid DNAs

1. Digest 5 µg of plasmid DNA using a restriction enzyme that has one site in the plasmid and leaves a convenient 5′ overhang for end-labeling. For example, if using pUC19 *Eco*RI is suitable.
2. Take two aliquots of 2 µg of the linearised plasmid DNA. One aliquot is "mock-methylated," i.e., treated in the same way as the other aliquot but with the omission of enzyme. Methylate the other aliquot using the CpG methylase (New England Biolabs 226S), which methylates all CpGs as directed by the manufacturer.
3. Assay if the methylation reaction has been successful by testing if the methylated DNA is now resistant to digestion by methylation-sensitive restriction enzymes such as *Hha*I or *Hpa*II. Perform reactions with and without enzyme following the manufacturer's instructions using about 30 ng DNA/reaction and analyze on a 1% agarose gel stained with ethidium bromide *(38)*. The "mock-methylated" DNA should be digested to completion by both *Hha*I and *Hpa*II, and the methylated DNA should be resistant to digestion by both enzymes.
4. Extract, precipitate, and resuspend the DNAs in 20 µL TE and measure the DNA concentration using standard procedures *(38)*.
5. Using standard procedures *(38)* end-label 600 ng of both the unmethylated and methylated linearized plasmids using the Klenow enzyme and appropriate labeled

Isolation of CpG Islands

and unlabeled nucleotides. For example, if the plasmid has been linearized with *Eco*RI, which leaves a 5′-AATT-3′ overhang, use $[\alpha]^{32}$P dATP and dTTP.

6. To eliminate unincorporated radioactivity ethanol precipitate the DNAs, centrifuge at 14 K (full speed) in a microfuge for 10 min and wash the DNA pellets twice with 70% ethanol using standard procedures *(38)*. Dry the DNA pellets either by air-drying or under vacuum and resuspend each in 600 µL of MBD buffer/ 0.1 *M* NaCl. Monitor each using a handheld Geiger counter to check for successful end-labeling. This amount is sufficient for 6 column runs and can be stored at 4°C for 2–4 wk.

3.3.2. Calibration of a MBD Column Using the End-Labeled Plasmid DNAs

1. Mix together 100 ng (100 µL) of each of the end-labeled unmethylated and completely methylated plasmid DNAs.
2. Load the DNA mixture onto a 1 mL MBD column, wash with MBD buffer/ 0.1 *M* NaCl up to 5 mL. Then wash with 5 mL of MBD buffer/0.4 *M* NaCl followed by a 40 mL linear salt gradient to 1 *M* NaCl (i.e., increase the concentration of NaCl from 0.4 *M* to 1 *M* over 40 mL). Finally wash with 5 mL of MBD buffer/1 *M* NaCl. Collect 1-mL fractions in either 5 mL or 1.5 mL tubes as convenient.
3. Count the radioactivity in each fraction. The radioactivity should elute from the column in two peaks during the linear gradient part of the run. Typically, the first peak will elute at about 0.5 *M* NaCl and the second peak at about 0.8 *M* NaCl.
4. From peak fractions remove 400 µL aliquots. Ethanol precipitate and resuspend DNA from these in 10 µL of TE using standard procedures *(38)*.
5. Determine the methylation status of these DNA samples by restriction-enzyme analysis as described in **Subheading 3.3.1., step 3**, using 3 µL of the test DNA/reaction. After running the analytical gel, dry it down, and expose it to X-ray film to visualize the end-labeled DNA fragments. The DNA in the first peak should be digested by both *Hpa*II and *Hha*I, showing that it is unmethylated. The DNA in the second peak should be resistant to digestion by both enzymes, showing that it is completely methylated.

3.3.3. Determination of the NaCl Concentration at Which Only Methylated DNA Binds to the MBD Column

In some applications, for example CpG island library construction, it is useful to load DNA onto the MBD column such that unmethylated DNA, which includes the CpG island DNA, remains in the flow-through. The NaCl concentration at which this happens can be determined using end-labeled unmethylated and "partially methylated" plasmid DNAs (*see* **Note 14**). These are loaded on the MBD column, this time individually, in MBD buffer containing various test NaCl concentrations to identify the highest at which the unmethylated plasmid does not bind to the column but the partially methylated plasmid does.

1. Load 100 ng of the end-labeled, unmethylated test plasmid onto the MBD column in 500 µL of MBD buffer/0.5 M NaCl. Wash the column with 9.8 mL of the MBD buffer/0.5 M NaCl followed by 10 mL of MBD buffer/1 M NaCl. Collect 1 mL fractions in 1.5 mL microfuge tubes or 5-mL tubes as convenient.
2. Count all the collected fractions for radioactivity to determine where the DNA elutes.
3. Reequilibrate the MBD column by washing with 10 mL of MBD buffer/ 0.5 M NaCl.
4. Repeat **steps 1–3** with the partially methylated plasmid.
5. Repeat **steps 1–4** varying the NaCl concentration of the MBD buffer in which the DNA is loaded onto the column in increments of 0.05 M to determine the highest at which at which the unmethylated DNA elutes in the loading buffer and the partially methylated DNA binds and is eluted by MBD buffer/1 M NaCl. Between each round of testing reequilibate the MBD column using MBD buffer containing the appropriate NaCl concentration.

3.4. Preparation of DNAs

3.4.1. Preparation and Digestion of Genomic DNA

1. Isolate genomic DNA from either blood or tissue samples using standard procedures *(38)*.
2. Digest the DNA to completion with a restriction enzyme whose recognition sequence is found infrequently within CpG island DNA but frequently elsewhere in the genome, such as *Mse*I, as directed by the manufacturer.
3. Following digestion, extract the DNA once with phenol, once with phenol/chloroform, once with chloroform, and ethanol-precipitate using standard procedures *(38)*. Wash the DNA pellet with 70% ethanol, dry either by air-drying or by under vacuum and resuspend it in 250 µL of MBD buffer containing between 0–0.6 M NaCl. The exact NaCl concentration is determined by the procedure to be carried out and also depends on the calibration results of the MBD column to be used. Store at –20°C until required.
4. In most applications, for example the preparation of CpG island libraries, the genomic DNA is methylated. Use the CpG methylase (New England Biolabs, 226S) as described in **Subheading 3.3.1.** to do this.

3.4.2. Preparation and Digestion of Cloned DNA

1. Prepare cloned DNAs using standard procedures *(38)* or using commercially available kits. As a final step, purify using CsCl-gradient *(38)*.
2. Digest DNA as described for genomic DNA in **Subheading 3.4.1.**
3. Carry out methylation of cloned DNAs as described for the test plasmids in **Subheading 3.3.1.**

3.5 Applications of MBD Columns

When using MBD columns, it is either fractions containing fragments that do not bind tightly, containing unmethylated DNA, or fractions containing fragments that bind tightly, containing methylated DNA, that are selected. Whether the former or the latter is determined by the aim of the experiment and/or stage of the protocol being carried out. Modify the protocols described in **Subheadings 3.2.** and **3.4.** as required. Here only the steps necessary required for each application are outlined.

3.5.1. Preparation of CpG Island Libraries from Genomic DNAs

1. Genomic DNA, digested as described in **Subheading 3.4.1.**, is loaded onto the MBD column and the fraction that does not bind tightly is selected (the stripped DNA). This is most easily done by loading the DNA in and washing with MBD buffer which is at the NaCl concentration determined in the calibration step (**Subheading 3.3.3.**) at which unmethylated DNA does not bind to the column but methylated DNA does.
2. Methylate the stripped DNA at all CpGs (*see* **Subheading 3.4.1.**).
3. Load the methylated stripped DNA onto the MBD column and this time select the fraction that binds tightly and elutes at the NaCl concentration determined in the calibration protocol (**Subheading 3.3.2.**) at which the heavily methylated test plasmid DNA elutes.
4. Isolate DNA from these fractions by precipitating with ethanol using standard procedures *(38)*. Include 20 µg of glycogen (1 µL of a 20 mg/mL solution Boehringher Mannheim 901 393) as a carrier. Resuspend the DNA in about 20 µL of TE. Clone this DNA into a suitable cloning vector to make a library (*see* **Subheading 4.5.**).
5. Analyze clones to check that they are derived from CpG islands (*see* **Subheading 4.5.**).

3.5.2. Isolation of CpG Islands from Large Genomic Clones

1. Digested and methylated cloned DNA, prepared as described in **Subheading 3.4.2.** is loaded onto the MBD column.
2. The column is then washed to remove fragments that bind weakly. Fragments that bind tightly are eluted using wash buffer containing the NaCl concentration determined in the calibration protocol (**Subheading 3.3.2.**) at which the heavily methylated test plasmid DNA elutes.
3. Isolate DNA from these fractions by precipitating with ethanol using standard procedures *(38)*. Include 20 µg of glycogen (1 µL of a 20 mg/mL solution Boehringher Mannheim 901 393) as a carrier. Resuspend the DNA in about 20 µL of TE. Clone this DNA using a suitable cloning vector (*see* **Subheading 4.5.**).
4. Analyze clones to check that they are derived from CpG islands (*see* **Subheading 4.5.**).

4. Notes
4.1. Preparation of an MBD Column

1. The plasmid pET6HMBD in the *E. coli* strain XL1-BLUE can be obtained by writing to S. H. Cross or Professor A. P. Bird, ICMB, Edinburgh University, King's Buildings, Mayfield Road, Edinburgh EH9 3JR. For expression of the recombinant protein HMBD pET6HMBD should be transformed into the *E. coli* strain BL21 (DE3) pLysS (F$^-$ *ompT hsd*S_B(r_B-m_B^-) *gal dcm* (DE3) pLysS (Novagen 69388-1). For some expression constructs, it has been found that expression levels tend to decrease if the same stock is used repeatedly. To avoid this, always use a freshly streaked plate from a frozen stock kept at –80°C. However, if expression problems persist, retransform pET6HMBD into BL21 (DE3) pLysS.
2. Buffers A, B, C, D, and E should be freshly prepared just before use.
3. In the initial analysis gel the induced HMBD protein may not be visible in samples 2, 3, and 4 because of the excess of bacterial proteins. However, after purification by cation exchange chromatography, the HMBD protein should be clearly visible and the dominant band present in the fractions eluted at high NaCl concentrations as most bacterial proteins elute in the flow through.
4. Alternative nickel-agarose resins to the Ni-NTA Superflow suggested here are available but be aware that some of these have to be charged before use (also *see* **Note 6**). Prepare the nickel-agarose resin to be used according to the manufacturer's directions. Failure of the HMBD protein to couple to nickel-agarose resin is most likely due to use of uncharged resin.
5. If a small amount of HMBD protein is present in the flow-through or wash fractions collected after the coupling it is likely that the capacity of the resin has been exceeded. Generally between 25 and 40 mg of HMBD is sufficient to saturate 1 mL of resin.

4.2. Basic Protocol for Running an MBD Column

6. Ideally, differentially methylated DNAs are separated by running MBD columns in conjunction with automated chromatography and fractionation systems such as the FPLC System (Pharmacia 18-1035-00) or Gradifrac System (Pharmacia 18-1993-01) with the resin packed into an HR5/5 column (Pharmacia 18-0382-01) so that flow rates and elution gradients can easily be controlled. Generally a flow rate of 1 mL/min is used. However, if such a system is not available, MBD columns can be made using a small disposable plastic chromatography column (for example Poly-Prep chromatography column Bio-Rad 731-1550) and run under gravity flow. In this case, I would suggest using Ni-NTA agarose (Qiagen 30210) rather than Ni-NTA Superflow as it is cheaper, has similar binding capacity, and the superior mechanical stability and flow characteristics of the Superflow resin are not required for gravity-flow applications.
7. MBD buffers should be freshly prepared just before use.
8. MBD columns should be calibrated before use with test plasmid DNAs containing known numbers of methyl-CpGs (*see* **Subheading 3.3.**).

9. In cases where DNAs are loaded onto the MBD column at NaCl concentrations higher than 0.1 M, use MBD buffer containing the appropriate NaCl concentration, instead of MBD buffer/0.1 M NaCl, when equilibrating the column.
10. This is only the basic procedure and should be adjusted and modified according to requirements. Firstly, the NaCl concentration of the MBD buffer in which DNAs are loaded onto the column can be adjusted. DNA binds to the MBD column, irrespective of methylation status, if loaded in MBD buffer/0.1 M NaCl. This is probably because the HMBD protein is very basic *(21)*. However, if DNAs are loaded in MBD buffer containing about 0.5 M NaCl it has been found that unmethylated DNA does not bind and remains in the flow-through but methylated DNA still does bind *(27,41)*. The highest molarity at which this happens will vary from column to column depending on the amount of coupled HMBD and should be determined as described in **Subheading 3.3.3**. Secondly, choose whether to elute bound DNAs by increasing the NaCl concentration of the wash buffer in steps, as a linear gradient or by a combination of the two. If using step-wise elutions, wash the column with 5 columns volumes of buffer at each step. Generally when eluting bound DNAs with linear gradients, the more shallow a gradient chosen the better the resolution. When using a 1 mL column, a linear gradient of 0.5–1 M over 30 mL has been found to give good separation of methylated DNAs *(27)*. If using step-wise elution, increase the concentration of NaCl by 100 mM NaCl for each step, which also results in good separation (S. H. Cross, unpublished observations).
11. MBD columns are stable for at least 6 mo if kept at 4°C and can be reused many times. Do not allow MBD columns to dry out.

4.3. Calibrating an MBD Column

12. The NaCl concentration at which a fragment elutes from the MBD column is determined principally by the total number of methylated CpGs it contains, rather than the number of CpGs per unit length *(21)*. Therefore it can be assumed that methylated CpG islands will elute at the same NaCl concentration as the heavily methylated plasmid DNA used to calibrate the column.
13. If the methylated plasmid is still susceptible to digestion by methylation-sensitive restriction enzymes repeat the methylation reaction. It is often necessary to do at least two rounds of methylation. Between each round extract and precipitate the DNA using standard procedures *(38)*.
14. When calibrating an MBD column plasmids containing a range of different numbers of methyl-CpGs can be used to refine where DNA fragments containing different numbers of methyl-CpGs can be expected to elute from the column. They can also be used for assaying the highest NaCl concentration at which unmethylated DNA does not bind to the MBD column, but methylated DNA still does and are used in **Subheading 3.3.3**. Such test plasmids can be prepared using methylase enzymes that modify CpGs within certain sequence contexts. For example *Hha*I and *Hpa*II methylases (New England Biolabs 217S and 214S) methylate CpGs within the sequence contexts CCGG and GCGC, respectively.

In the case of pUC19, use of these enzymes together would yield a plasmid containing 30 methylated CpGs.
15. Here only the modifications required for calibration are detailed. Refer to **Subheading 3.2.** for the basic procedure that should be followed when running an MBD column.
16. If it is not possible to increase the NaCl concentration using a linear gradient increase the NaCl concentration in steps of 0.1 M NaCl from 0.4 M to 1 M.
17. To avoid loss of small DNA fragments during drying of the analytical gel, place it on DE81 paper (Whatman 3658 915), which is then placed on two sheets of 3MM paper (Whatman 3030 917). Cover with clingfilm before drying down.

4.4. Preparation of DNAs

18. Do not use placental or sperm DNA as a source of genomic DNA, because satellite sequences, which contain many CpGs that are not associated with genes, are often undermethylated in these tissues *(42)*. Established cell-lines are also generally not a good choice because those CpG islands associated with genes having a tissue-restricted pattern of expression are often methylated *(8)*.
19. Generally 100–200 µg of genomic DNA is sufficient for construction of a CpG island library. When isolating CpG island DNA from a genomic clone, generally 10–20 µg of DNA is required.
20. *Mse*I recognizes the sequence TTAA that is predicted to occur, on average, every 1000 bp within CpG islands and every 150–200 bp elsewhere *(21)*. However, the dinucleotide TA is found less frequently than expected in the genome, for reasons that are not understood, so that *Mse*I sites occur less frequently than they are predicted to. This has the advantage that the chance of an *Mse*I site occurring within a CpG island is reduced. On the other hand, the size of other genomic fragments is larger than expected, but this does not matter because of the low frequency of CpG in the genome. Following *Mse*I digestion up to two-thirds of CpG islands are left intact, whereas other sequences are found on small fragments containing on average, 1–5 methylated CpGs *(21)*. Other restriction enzymes with a 4 bp recognition site containing only Ts and As, such as *Tsp509* I, which recognizes the sequence AATT, could be used, although sites for such enzymes may be found more frequently within CpG islands.
21. *Mse*I is a convenient enzyme to use because *Mse*I fragments can be cloned into the *Nde*I site of the pGEM®-5Zf(+/–) cloning vectors (Promega P2241 and P2351) (*see* **Note 30** for discussion of cloning of purified CpG island fragments).
22. To methylate 100–200 µg of genomic DNA, perform a 500 µL reaction. To monitor, remove two aliquots of 10 µL of the reaction mix before and after the addition of the enzyme. To these add 1 µg of linearized plasmid DNA such as *Eco*RI-digested pUC19. After incubation, analyze these as described in **Subheading 3.3.1., step 3** using 3 µL/restriction digest. Successful methylation of the plasmid DNA, as indicated by resistance to digestion by *Hpa*II and *Hha*I shows that the methylation of the genomic DNA has also gone to completion. If not, do another round of methylation (*see* **Subheading 4.3., Note 2**).

23. When fractionating CpG islands from genomic clones (for example cosmids, PACs and BACs) it is important to use CsCl-gradient purified DNAs. This is because any contaminating *E. coli* DNA present will co-purify with the CpG island DNA, which it resembles in sequence composition. This is also the case for yeast DNA. As it is extremely difficult to purify recombinant YACs free of contaminating yeast, DNA purification of CpG islands from such clones using an MBD column is not recommended. Alternative approaches to isolate CpG islands from YACs are probably more suitable *(43)*. As an alternative, CpG-island libraries could be used in an analogous way to cDNA selection *(44,45)* for direct selection of CpG islands from cloned DNA.

4.5. Applications of MBD Columns

24. When preparing CpG island libraries, the stripping step is done to remove any *Mse*I fragments that contain a cluster of methylated CpGs and that would contaminate the final library. For efficient removal of such highly methylated fragments, up to three rounds of column chromatography are necessary. Monitor at each round for removal of such sequences (*see* **Note 28**).
25. In some cases, it is DNA fragments that are heavily methylated in genomic DNA that are of interest *(35,37)*. To purify such fragments, load genomic DNA on the MBD column as described in **Subheading 3.5.1.**, wash to remove lightly methylated fragments, and then wash at high salt to elute heavily methylated fragments. Perform at least two rounds of binding so that this fraction is purified away from unmethylated fragments efficiently. Between each round, dilute with MBD buffer so that the NaCl concentration is reduced to that at which unmethylated DNA does not bind to the column and methylated DNA does, as determined in the calibration protocol **Subheading 3.3.3.** Monitor at each round for removal of such sequences (*see* **Note 28**).
26. If preparing a CpG island library with limiting amounts of DNA, for example if using sorted chromosomes *(27)*, catch-linkers can be attached to the DNA in the stripped fraction allowing it to be amplified by PCR. Methylation patterns are erased by doing this but this does not matter as the next step of the protocol is to methylate all CpGs.
27. The stripped fraction contains unmethylated CpG island fragments, which contain many unmethylated CpGs, along with fragments from elsewhere in the genome that contain only a few CpGs, most of which are methylated. By methylating all the nonmethylated CpGs in this fraction, the affinity of CpG island fragments for binding to the MBD column is changed from being weak to being strong because of the high number of CpGs present. On the other hand, the affinity of other genomic fragments is unaltered as they contain only a few CpGs, most of which are methylated already. Therefore, by loading the methylated stripped DNA onto the MBD column and now selecting fragments that elute at high salt (as determined in the calibration protocol, **Subheading 3.3.3.**), a fraction that is highly enriched for CpG islands is obtained. Generally at least two rounds of binding are required to yield a preparation of highly purified

CpG islands. Between each round dilute with MBD buffer so that the NaCl concentration is reduced to that at which unmethylated DNA does, not bind to the column and methylated DNA does as determined in the calibration protocol, **Subheading 3.3.3**. Monitor as described in **Note 28**.

28. Assay for the presence or absence of representative *Mse*I fragments at the different stages of the purification procedure to evaluate if additional rounds of column binding are required. *Mse*I fragments chosen for this should be derived from: 1) CpG island DNA, 2) bulk genomic DNA, and 3) GC-rich heavily methylated DNA. This can be done by performing PCR reactions on aliquots of the various fractions using primers that amplify sequences from each type of fragment. Such analysis was used during the construction of a human CpG island library *(21)*.

29. When isolating CpG islands from genomic clones, the stripping step that removes GC-rich, methylated DNA cannot be carried out because on cloning the native methylation pattern is erased. Any such fragments present in the clones will co-purify with the CpG islands. The genomic methylation status of clones, and therefore whether or they are genuine CpG islands, can be determined by doing Southern blot analysis, as described in **Note 32**. However, bear in mind that in some cases CpG islands are methylated as discussed in the Introduction.

30. As mentioned in **Note 21**, *Mse*I fragments can be cloned into the *Nde*I site of plasmid vectors such as pGEM®-5Zf(–/+) (Promega P2241 and P2351). This is because *Mse*I and *Nde*I produce compatible cohesive ends, which are therefore compatible for ligation. As the cloning site is destroyed, the best way to examine clone inserts is to amplify them by PCR using primers flanking the cloning site. Dephosphlorylate the linearized vector before use to reduce background, using standard procedures *(38)*. Use standard techniques for both ligation of the CpG island fraction into the vector and transformation *(38)*.

31. The bacterial strain chosen for transformation should be one that does not restrict methylated DNA, such as SURE (Stratagene 200294).

32. Analysis of potential CpG island clones should be carried out to determine if they are derived from *bone fide* CpG islands. The sequence of the clones would be expected to have a GC-content in excess of 50% and to contain close to the expected number of CpGs. The clones would also be expected to be derived from unmethylated genomic sequences. For CpG island libraries prepared from genomic DNA, analysis of 20 clones is usually sufficient to give a good idea of the veracity of a CpG-island library. Suggested tests are:
    a. Test clones for the presence of *Bst*UI sites. This restriction enzyme has the recognition sequence CGCG, which occurs about 1/100 bp in CpG island DNA and about 1/10 kb in non-CpG island DNA. If a clone contains a *Bst*UI site this is a good indication that it is derived from a CpG island. This is an easy and reliable way of quickly judging if clones are from CpG islands.
    b. Sequence clones to determine if they have the sequence characteristics of CpG islands. Expect a G+C content of greater than 50% and close to the expected number of CpGs, as predicted from base composition. Mammalian genomic DNA has a G+C content of about 40% and contains only about 25% of the

Isolation of CpG Islands

      expected number of CpGs as predicted from base composition. An easy way to visualize this data is to plot a graph with base composition on the x axis and CpG observed/expected values on the y axis, see *(21)* for an example.

   c. Search sequence databases to check that clones are derived from the expected source species, from known CpG islands and that there is no major contamination, from bacterial sources for example.

   d. Determine whether the clones are derived from unmethylated DNA in the genome. Use clones that do not contain repeats (*see* e) to probe Southern blots of genomic DNA that has been digested with *Mse*I alone and *Mse*I and a methylation-sensitive restriction enzyme such as *Bst*UI or *Hpa*II, using standard procedures *(38)*. If the clone is derived from an unmethylated CpG island, the *Mse*I fragments should be cleaved by the methylation-sensitive enzymes.

   e. For use in (d) and for CpG island libraries test if clones contain repeated sequences by hybridising colonies with total genomic DNA. One characteristic of CpG island libraries is that a low proportion of the clones (about 10%) contain highly repeated sequences *(21)*.

   f. For CpG island libraries assay the proportion of clones that contain ribosomal sequences (if the starting DNA contained ribosomal sequences). Transcribed ribosomal DNA has the same sequence characteristics as CpG islands and is also unmethylated in genomic DNA *(35,46)*. Therefore it co-purifies with CpG island DNA and should be present in the final CpG island fraction.

33. Especially when isolating CpG islands from genomic clones, it should be remembered that they are not found at the 5' end of all genes, notable exceptions being many genes with a tissue-restricted pattern of expression *(5,6)*. Therefore, in positional cloning projects for example, other methods such as exon trapping and cDNA selection should be used to identify such genes *(44,45,47)*. However, the method described here does have the advantage that it depends only on sequence composition and is unaffected by gene expression patterns.

34. CpG islands are useful gene markers because there is only one CpG island/gene, they co-localize with the 5' end of the transcript, include promoter sequences, and as they are usually single-copy, they can be used to map genes and to isolate full-length cDNAs.

*The estimate of the number of CpG islands present in the human genome was recently modified downwards to 34,200 *(48)*. This figure is reasonably close to the 28,890 potential CpG islands that have been identified so far in the draft human genomic sequence *(49)*. The estimate of the proportion of genes that have a CpG island remains unchanged.

References

1. Antequera, F. and Bird, A. (1993) CpG islands, in *DNA Methylation: Molecular Biology and Biological Significance* (Jost, J. P. and Saluz, H. P., eds.), Birkhauser Verlag, Basel, Switzerland, pp. 169–185.
2. Cross, S. H. and Bird, A. P. (1995) CpG islands and genes. *Curr. Opin. Genet. Dev.* **5,** 309–314.

3. Tazi, J. and Bird, A. (1990) Alternative chromatin structure at CpG islands. *Cell* **60,** 909–920.
4. Antequera, F. and Bird, A. (1993) Number of CpG islands and genes in human and mouse. *Proc. Natl. Acad. Sci. USA* **90,** 11,995–11,999.
5. Gardiner-Garden, M. and Frommer, M. (1987) CpG islands in vertebrate genomes. *J. Mol. Biol.* **196,** 261–282.
6. Larsen, F., Gunderson, G., Lopez, R., and Prydz, H. (1992) CpG islands as gene markers in the human genome. *Genomics* **13,** 1095–1107.
7. Riggs, A. D. and Pfeifer, G. P. (1992) X-chromosome inactivation and cell memory. *Trends Genet.* **8,** 169–174.
8. Antequera, F., Boyes, J., and Bird, A. (1990) High levels of *de novo* methylation and altered chromatin structure at CpG islands in cell-lines. *Cell* **62,** 503–514.
9. Tilghman, S. M. (1999) The sins of the fathers and mothers: genomic imprinting in mammalian development. *Cell* **96,** 185–193.
10. Greger, V., Passarge, E., Höpping, W., Messmer, E., and Horsthemke, B. (1989) Epigenetic changes may contribute to the formation and spontaneous regression of retinoblastoma. *Hum. Genet.* **83,** 155–158.
11. Issa, J-P. J., Ottaviano, Y. L., Celano, P., Hamilton, S. R., Davidson, N. E., and Baylin, S. B. (1994) Methylation of the oestrogen receptor CpG island links ageing and neoplasia in human colon. *Nature Genet.* **7,** 536–540.
12. Schmutte, C. and Jones, P. A. (1998) Involvement of DNA methylation in human carcinogenesis. *Biol. Chem.* **379,** 377–388.
13. Macleod, D., Charlton, J., Mullins, J., and Bird, A. (1994) Sp1 sites in the mouse *Aprt* gene promoter are required to prevent methylation of the CpG island. *Genes Dev.* **8,** 2282–2292.
14. Brandeis, M., Frank, D., Keshet, I., Siegfried, Z., Mendelsohn, A., Nemes, A., et al. (1994) Sp1 elements protect a CpG island from *de novo* methylation. *Nature* **371,** 435–438.
15. Tykocinski, M. L. and Max, E. E. (1984) CG dinucleotide clusters in MHC genes and in 5′ demethylated genes. *Nucleic Acids Res.* **12,** 4385–4396.
16. Stöger, R., Kubicka, P, Liu, C. G., Kafri, T., Razin, A., Cedar, H., and Barlow, D. P. (1993) Maternal-specific methylation of the imprinted mouse *Igf2r* locus identifies the expressed locus as carrying the imprinting signal. *Cell* **73,** 61–71.
17. Macleod, D., Ali, R. R., and Bird, A. (1998) An alternative promoter in the mouse major histocompatibility complex class II I-Aβ gene: implications for the origin of CpG islands. *Mol. Cell Biol.* **18,** 4433–4443.
18. Wutz, A., Smrzka, O. W., Schweifer, N., Schellander, K., Wagner, E. F., and Barlow, D. P. (1998) Imprinted expression of the Igf2r gene depends on an intronic CpG island. *Nature* **389,** 745–749.
19. Delgado, S., Gómez, M., Bird, A., and Antequera, F. (1998) Initiation of DNA replication at CpG islands in mammalian chromosomes. *EMBO J.* **17,** 2426–2435.
20. Bickmore, W. A. and Bird A. P. (1992) Use of restriction endonucleases to detect and isolate genes from mammalian cells. *Methods Enzymol.* **216,** 224–245.

21. Cross, S. H., Charlton, J. A., Nan, X., and Bird, A. P. (1994) Purification of CpG islands using a methylated DNA binding column. *Nature Genet.* **6,** 236–244.
22. Lewis, J. D., Meehan, R. R., Henzel, W. J., Maurer-Fogy, I., Jeppesen, P., Klein, F., and Bird, A. (1992) Purification, sequence and cellular localisation of a novel chromosomal protein that binds to methylated DNA. *Cell* **69,** 905–914.
23. Nan, X., Meehan, R. R., and Bird, A. (1993) Dissection of the methyl-CpG binding domain from the chromosomal protein MeCP2. *Nucleic Acids Res.* **21,** 4886–4892.
24. Nan, X. Campoy, J., and Bird, A. (1997) MeCP2 is a transcriptional repressor with abundant binding sites in genomic chromatin. *Cell* **88,** 471–481.
25. Nan, X., Ng, H., Johnson, C. A., Laherty, C. D., Turner, B. M., Eisenman, R. N., and Bird, A. (1998) Transcriptional repression by the methyl-CpG-binding protein MeCP2 involves a histone deacetylase complex. *Nature* **393,** 386–389.
26. Jones, P. L., Veenstra, G. J. C., Wade, P. A., Vermaak, D., Kass, S. U., Landsberger, N., et al. (1998) Methylated DNA and MeCP2 recruit histone deacetylase to repress transcription. *Nature Genet.* **19,** 187–191.
27. Cross, S. H., Lee, M., Clark, V. H., Craig, J. M., Bird, A. P., and Bickmore, W. A. (1997) The chromosomal distribution of CpG islands in the mouse: evidence for genome scrambling in the rodent lineage. *Genomics* **40,** 454–461.
28. McQueen, H. A., Fantes, J., Cross, S. H., Clark, V. H., Archibald, A. L., and Bird, A. P. (1996) CpG islands of chicken are concentrated on microchromosomes. *Nature Genet.* **12,** 321–324.
29. McQueen, H. A., Clark, V. H., Bird, A. P., Yerle, M., and Archibald, A. L. (1997) CpG islands of the pig. *Genome Res.* **7,** 924–931.
30. McQueen, H. A., Siriaco, G., and Bird, A. P. (1998) Chicken microchromosomes are hyperacetylated, early replicating, and gene rich. *Genome Res.* **8,** 621–630.
31. Craig, J. M. and Bickmore, W. A. (1994) The distribution of CpG islands in mammalian chromosomes. *Nature Genet.* **7,** 376–382.
32. Watanabe, T., Inoue, S., Hiroi, H., Orimo, A., Kawashima, H., and Muramatsu, M. (1998) Isolation of estrogen-responsive genes with a CpG island library. *Mol. Cell Biol.* **18,** 442–449.
33. Cross, S. H., Clark, V. H., and Bird, A. P. (1999) Isolation of CpG islands from large genomic clones. *Nucleic Acids Res.* **27,** 2099–2107.
34. Cross, S. H., Clark, V. H., Simmen, M. W., Bickmore, W. A., Maroon, H., Langford, C. F., et al. (1999) Preparation and characterisation of CpG islands libraries from human chromosomes 18 and 22: landmarks for novel genes. *Mamm. Gen.* **11,** 373–383.
35. Brock, G. J. R. and Bird, A. (1997) Mosaic methylation of the repeat unit of the human ribosomal RNA genes. *Hum. Mol. Genet.* **6,** 451–456.
36. Huang, T. H., Perry, M. R., and Laux, D. E. (1999) Methylation profiling of CpG islands in human breast cancer cells. *Hum. Mol. Genet.* **8,** 459–470.
37. Shiraishi, M., Chuu, Y., and Sekiya, T. (1999) Isolation of DNA fragments associated with methylated CpG islands in human adenocarcinomas of the lung using

a methylated DNA binding column and denaturing gradient gel electrophoresis. *Proc. Natl. Acad. Sci. USA* **96,** 2913–2918.
38. Sambrook, J., Fritsch, E. F., and Maniatis, T. (eds.) (1989) *Molecular Cloning. A Laboratory Manual.* 2nd ed. Cold Spring Harbor Laboratory Press, Cold Spring Harbor, NY.
39. Studier, F. W., Rosenberg, A. H., Dunn, J. J., and Dubendorff, J. W. (1990) Use of T7 RNA polymerase to direct expression of cloned genes. *Methods Enzymol.* **185,** 60–89.
40. Bradford, M. (1976) A rapid and sensitive method for the quantitation of microgram quantities of protein utilising the principle of protein dye binding. *Anal. Biochem.* **72,** 248–254.
41. John, R. M. and Cross, S. H. (1997) Gene detection by the identification of CpG islands, in *Genome Analysis: A Laboratory Manual, vol. 2 Detecting Genes* (Birren, B., Green, E. D., Klapholz, S., Myers, R. M., and Roskams, J., eds.), Cold Spring Harbor Laboratory Press, Cold Spring Harbor, NY, pp. 217–285.
42. Sanford, J., Chapman, V. M., and Rossant, J. (1985) DNA methylation in extra-embryonic lineages of mammals. *Trends Genet.* **1,** 89–93.
43. Valdes J. M., Tagle, D. A., and Collins F. S. (1994) Island rescue PCR: a rapid and efficient method for isolating transcribed sequences from yeast artificial chromosomes and cosmids. *Proc. Nat. Acad. Sci. USA* **91,** 5377–5381.
44. Parimoo, S., Patanjali, S. R., Shukla, H., Chaplin, D. D., and Weissman, S. M. (1991) cDNA selection: efficient PCR approach for the selection of cDNAs encoded in large chromosomal DNA fragments. *Proc. Natl. Acad. Sci. USA* **88,** 9623–9627.
45. Lovett, M., Kere, J., and Hinton, L. M. (1991) Direct selection: a method for the isolation of cDNAs encoded by large genomic regions. *Proc. Natl. Acad. Sci. USA* **88,** 9628–9632.
46. Bird, A. P. and Taggart, M. H. (1980) Variable patterns of total DNA and rDNA methylation in animals. *Nucleic Acids Res.* **8,** 1485–1497.
47. Buckler, A. J., Chang, D. D., Graw, S. L., Brook, J. D., Haber, D. A., Sharp P. A., and Housman, D. E. (1991) Exon amplification: A strategy to isolate mammalian genes based on RNA splicing. *Proc. Natl. Acad. Sci. USA* **88,** 4005–4009.
48. Ewing, B., and Green, P. (2000) Analysis of expressed sequence tags indicates 35,000 human genes. *Nat. Genet.* **25,** 232–234.
49. International Human Genome Sequencing Consortium (2001) Initial sequencing and analysis of the human genome. *Nature* **409,** 860–921.

11

Purification of MeCP2-Containing Deacetylase from *Xenopus laevis*

Peter L. Jones, Paul A. Wade, and Alan P. Wolffe

1. Introduction

DNA methylation has long been associated with stable transcriptional silencing and a repressive chromatin structure (review **refs. *1*,*2***). Differential methylation is associated with imprinting, carcinogenesis, silencing of repetitive DNA, and allows for differentiating cells to efficiently shut off unnecessary genes. In vertebrates, where 60–90% of genomic CpG dinucleotides are methylated, methylation-dependent repression is vital for proper embryonic development *(3)*. Microinjection experiments using methylated DNA templates implicate chromatin structure as an underlying mechanism of methylation-dependent silencing *(4,5)*. Methyl-specific transcriptional repression requires chromatin assembly, and can be partially relieved by the histone deacetylase inhibitor Trichostatin A. In addition, two proteins have been identified, MeCP1 *(6)* and MeCP2 *(7)*, that specifically bind to methylated DNA and mediate transcriptional repression. MeCP1 is a relatively uncharacterized complex that requires at least 12 symmetrical methyl-CpGs for DNA binding *(6)*. MeCP2 is a single polypeptide containing a methyl-binding domain capable of binding a single methyl-CpG, and a transcriptional repression domain *(8)*. Recently MeCP2 was shown to interact with the Sin3 corepressor and histone deacetylase *(9,10)*. Changes in the acetylation state of the core histone tails correlates with changes in transcription (reviewed in **refs. *11,12***), and several transcriptional repression complexes containing histone deacetylases have recently been described *(9,13,14)*. These data provide a direct link between methyl-dependent transcriptional repression and the modification of chromatin structure. Here,

From: *Methods in Molecular Biology, vol. 200: DNA Methylation Protocols*
Edited by: K. I. Mills and B. H. Ramsahoye © Humana Press Inc., Totowa, NJ

we describe techniques for purifying the MeCP2-contining histone deacetylase complex from *Xenopus laevis* oocytes.

Purification schemes for DNA-binding proteins often utilize the specific DNA-binding site sequences for the protein of interest in a DNA-binding type assay. These assays use protein binding to a radiolabeled nucleic acid as the readout, however, Southwestern analysis allows for the separation of multiple peptides in a sample that may bind to the same probe. Southwestern analysis is based on the ability of many proteins to be denatured with guanidine hydrochloride (G-HCl) and renatured such that the protein, or a portion thereof, refolds such that the DNA binding activity is retained *(15,16)* (*see* **Note 1**). Protein extracts are separated by size on sodium dodecyl sulfate-polyacrylamide gel electroporesis (SDS-PAGE), immobilized to a membrane, denatured with G-HCl followed by renaturation, and then hybridized with a radiolabeled nucleic acid probe. MeCP2 was originally characterized from rat by the Southwestern technique *(7)* and has recently been cloned from *X. laevis* (migrating at a molecular weight of 87 kDa on SDS-PAGE) *(9)*. Thus, MeCP2 can be very accurately monitored by southwestern assay as the 87 kDa protein with binding specific to methylated DNA (**Fig. 1A,B**).

Due to the association of MeCP2 with a histone deacetylase complex, it is useful to follow histone deacetylase enzymatic activity. The purification of histone deacetylase complexes in large part depends on having a sensitive and reliable assay. We outline an assay that utilizes purified recombinant histone acetyltransferase and purified chicken erythrocyte histones to specifically label the desired histone to a high specific activity while retaining the native histone octamer. This assay allows for the monitoring of the histones deacetylase activity of the MeCP2 complex during complex purification from oocyte extract.

The major advantage in using *X. laevis* oocyte extracts is many chromatin components including MeCP2 are present in large quantities in storage forms, and these chromatin components can be extracted in low salt, preserving the integrity of the complexes *(17)*. Oocyte extracts prepared by this method contain robust histone deacetylase activity (**Fig. 2** and **refs.** *9* and *13*). Using the assays described to monitor MeCP2 and histone deacetylase activities through the following chromatography protocols, the MeCP2-histone deacetylase complex can be purified.

2. Materials
2.1. Southwestern Oligo Preparation
1. Complementary oligonucleotides were synthesized (Operon Technologies) either with (oligos 3 and 4) or without (oligos 1 and 2) 5-methylcytosine (M) at each CpG residue for the following sequences *(7)*:

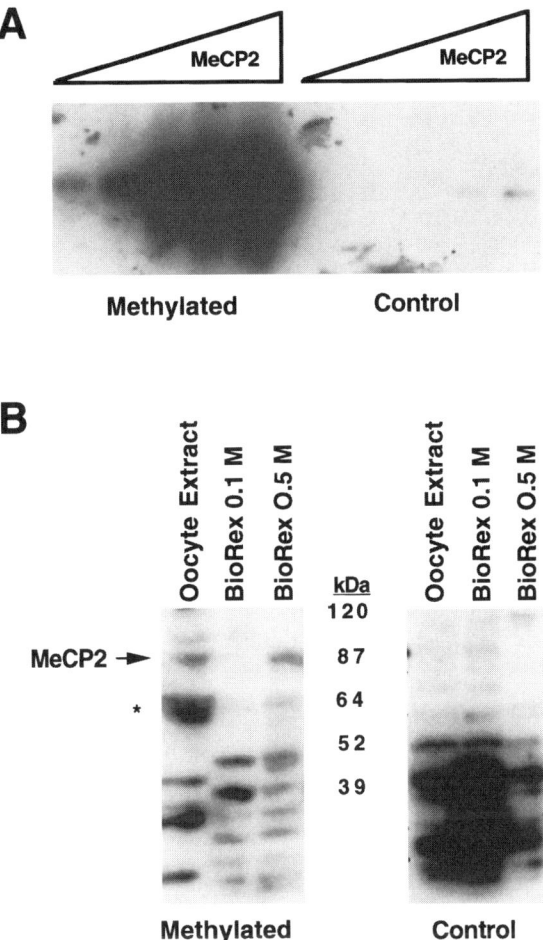

Fig. 1. Southwestern-blot analysis of (A) recombinant *X. laevis* MeCP2 and (B) endogenous *X. laevis* MeCP2 shows the preference of MeCP2 binding for methylated DNA. (A) Increasing amounts of recombinant xMeCP2 (75–900 ng) hybridized with either a methylated (left) or control unmethylated (right) probe. (B) Fractionation of oocyte extract over BioRex70 resin hybridized with either a methylated (left) or control unmethylated (right) probe. MeCP2 runs at 87 kDa. The asterik indicates a degradation product.

Oligo 1 GATCCGACGACGACGACGACGACGACGACGACGACGACGATC
Oligo 2 GATCGTCGTCGTCGTCGTCGTCGTCGTCGTCGTCGTCGGATC
Oligo 3 GATCMGAMGAMGAMGAMGAMGAMGAMGAMGAMGAMGAM
GATC
Oligo 4 GATMGTMGTMGTMGTMGTMGTMGTMGTMGTMGTMGTMG
GATC

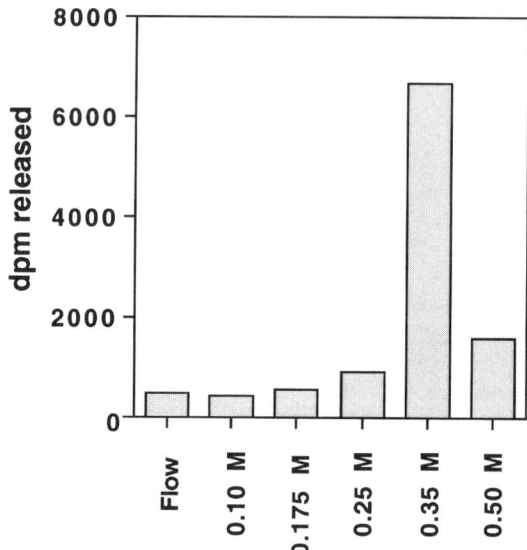

Fig. 2. Equal volumes of fractions from BioRex70 step-elutions of oocyte extract were assayed for deacetylase activity as described.

2. Elution buffer: 0.5 M ammonium acetate and 1 mM ethylenediaminetetraacetic acid (EDTA).
3. Kinase reagents: [^{32}P]γ-ATP (3000 Ci/mmol), and T4 polynucleotide kinase.

2.2. Southwestern Assay

1. 8% SDS gel with 4 % stacking gel.
2. 2X SDS-PAGE loading buffer.
3. Nitrocellulose membrane.
4. SW Transfer buffer: 25 mM Tris base and 190 mM glycine.
5. SW buffer: 20 mM HEPES, pH 7.9, 3 mM MgCl$_2$, 40 mM KCl, and 10 mM 2-mercaptoethanol.
6. SW Buffer + 6 M guanidine hydrochloride (G-HCl) (avoid contact with skin).
7. Blocking buffer: SW buffer + 2% nonfat dried milk.
8. Binding buffer: SW buffer, 25 µg/mL sonicated native *Escherichia coli* DNA, 2 µg/mL denatured *E. coli* DNA, and 0.1 % Triton X-100.
9. SW Washing buffer: SW buffer + 0.01% Triton X-100.

2.3. In Vitro Histone Acetylation

1. Purified chicken erythrocyte histones *(18)*.
2. Acetylation buffer: 1X = 25 mM Tris-HCl, pH 8.0, 100 mM NaCl, 0.1 mM EDTA, 0.2% PMSF, and 10% glycerol.
3. Recombinant Hat1p (*see* **Note 2**).

4. [^{3}H]Acetyl-coenzyme A (4.90 Ci/mmol) (Amersham Life Science).
5. Buffer A(200): 10 mM Tris-HCl, pH 8.0, 0.1 mM EDTA, and 200 mM NaCl.
6. Buffer A(2000): 10 mM Tris-HCl, pH 8.0, 0.1 mM EDTA, and 2 M NaCl.

2.4. Histone Deacetylase Assay

1. Deacetylase buffer 1X = 25 mM Tris-HCl, pH 8.0, 10% glycerol, 50 mM NaCl, and 1 mM EDTA.
2. [^{3}H] histones (*see* **Subheading 2.3.**).
3. Deacetylase stop solution: 0.1 M HCl and 0.16 M HAc.
4. Ethyl acetate.

2.5. Oocyte Extract Preparation

1. Female *Xenopus laevis*.
2. Dissection scissors and forceps.
3. SW-41Ti ultracentrifuge rotor and 12-mL tubes.
4. OR-2 buffer: 5 mM HEPES, pH 7.9, 1 mM Na$_2$(PO$_4$), 82.5 mM NaCl, 2.5 mM KCl, and 1 mM MgCl$_2$.
5. Extraction buffer: 20 mM HEPES, pH 7.5, 5 mM KCl, 1.5 mM MgCl$_2$, 1 mM EGTA, 10% glycerol, 10 mM β-glycerophosphate, 0.5 mM DTT, 1 mM PMSF, 2 µg/mL Pepstatin A, and 1 µg/mL leupeptin.

2.6. Chromatography

All buffers are at 4°C. DTT and protease inhibitors are added just prior to use. All buffers must be filtered through a 0.45-µm filter before use with the fast protein liquid chromatography (FPLC), Pharmacia, Biotech.

1. BioRex 70 resin 100-200 mesh (BioRad) equilibrated to Na$^+$ form.
2. Superose 6 HR 10/30 FPLC column (Pharmacia Biotech).
3. MonoQ sepharose HR 5/5 or HR 10/10 FPLC column (Pharmacia Biotech).
4. HiTrap Heparin 1 mL column (Pharmacia Biotech).
5. Buffer A(0): 20 mM HEPES, pH 7.5, 1.5 mM MgCl$_2$, 1 mM EGTA, 10 mM β-glycerophosphate, 10% glycerol, 0.5 mM dithiothreitol (DTT), 1 mM PMSF, 2 µg/mL pepstatin A, and 1 µg/mL leupeptin.
6. Buffer A(100). 100 mM NaCl, 20 mM HEPES, pH 7.5, 1.5 mM MgCl$_2$, 1 mM EGTA, 10 mM β-glycerophosphate, 10% glycerol, 0.5 mM DTT, 1 mM PMSF, 2 µg/mL pepstatin A, 1 µg/mL leupeptin.
7. Buffer A(500): 500 mM NaCl, 20 mM HEPES, pH 7.5, 1.5 mM MgCl$_2$, 1 mM EGTA, 10 mM β-glycerophosphate, 10% glycerol, 0.5 mM DTT, 1 mM PMSF, 2 µg/mL pepstatin A, 1 µg/mL leupeptin.
8. Buffer A(1000): 1 M NaCl, 20 mM HEPES, pH 7.5, 1.5 mM MgCl$_2$, 1 mM EGTA, 10 mM β-glycerophosphate, 10% glycerol, 0.5 mM DTT, 1 mM PMSF, 2 µg/mL pepstatin A, 1 µg/mL leupeptin.

3. Methods

3.1. Probe Preparation

1. Mix equimolar quantities of oligos 1 and 2 (20 µM final) and make to 1X TE + 0.1 M KCl in a 500 µL eppendorf tube.
2. Heat the oligos to 100°C for 2 min in a thermocyler and bring back to 30°C over 60 min, then immediately cool to 4°C.
3. Radiolabel using 4 pmol of duplex oligo (8 pmol ends) with 16 pmol ^{32}P-γ-ATP and T4 polynucleotide kinase for 60 min at 37°C.
4. Gel purify the kinase reactions on a 5% native polyacrylamide gel run at 200 V for 2 h.
5. Excised the labeled duplex from the gel with a razor blade and elute overnight 0.5 mL Elution buffer at 37°C with shaking.
6. Ethanol precipitated the probe and resuspended in 1X TE, pH 8.0. This purification step removes all unincorporated label as well as any single-strand oligonucleotides.

3.2 Southwestern Blotting

Dependent on the concentration of MeCP2 in the sample, a TCA precipitiation may have to be performed prior to running the SDS-PAGE (*see* **Note 3**).

1. The protein sample is made 1X in SDS-PAGE loading buffer and loaded directly onto an 8% SDS-PAGE gel with a 4% stacking gel and electrophoresed at 100 V in 4°C until the bromphenol blue reaches the bottom of the gel.
2. The proteins are transferred to nitrocellulose in 1 L SW Transfer buffer at 4°C in a Mini Trans-blot Transfer cell (BioRad) for 5 h at 350 mAmps.
3. The membranes are removed brom the cell and are soaked in SW buffer + 6 M G-HCl for 5 min with shaking at 4°C.
4. The filters are renatured by four twofold dilutions with SW buffer for 5 min each at 4°C with shaking.
5. Wash once with straight SW buffer.
6. Block the filters for 10 min at room temperature with SW Blocking buffer.
7. Rinsed the filters once in SW buffer.
8. Hybridization in Binding buffer with 2×10^5 CPM/mL labeled probe for 1 h at room temperature (*see* **Note 4**).
9. Washed the filters twice with SW Washing buffer for 5 min at room temperature, air-dry on 3MM paper, and exposed to film overnight with an intensifying screen.

3.3. In Vitro Histone Acetylation

1. Incubate one milligram core histones with 100 µg rHat1p and 100 µL ^{3}H acetyl coenzyme A (Amersham) in 1X Acetylation buffer in a final volume of 8.8 mL for 30 min at 37°C.

2. Chase by adding 100 nmoles cold acetyl coenzyme A and incubating for 30 min at 37°C.
3. Equilibrate a 1 mL BioRex70 column with Buffer A(200).
4. Load the acetylation reaction and allow it to enter the column by gravity.
5. Wash the column is with 5 mL Buffer A(200).
6. Eluted the histones with careful addition of 5 mL Buffer A(2000).
7. Collect the eluate in 0.5 mL fractions and assay by liquid scintillation (*see* **Note 5**).

3.4. Deacetylase Assay

1. Incubate a small volume of sample (1–10 µL) in a 200 µL reaction (made to 1X deacetylase buffer) with 1 µg acetylated histones, for 30 min at 30°C.
2. Add 50 µL stop solution.
3. Extracted the acetate from the reaction with 600 µL ethyl acetate.
4. Remove 450 µL of the organic phase to a liquid-scintillation vial.
5. Add liquid-scintillation fluid and count the samples in a liquid-scintillation counter.

3.5. Preparation of Oocyte Extracts (see Note 6)

1. Dissect ovaries from mature female *X. laevis* and rinse in OR-2 buffer.
2. Cut ovaries into small pieces and put into 50-mL conical tubes (15 mL ovary tissue per tube).
3. Rinse several times with OR-2.
4. Add fresh OR-2 is to a final volume of about 35 mL.
5. Add collagenase Type II (Sigma Chemical) to 0.75 mg/mL.
6. Place the ovaries on a platform shaker and agitate for 60–90 min until the oocytes are dispersed.
7. Wash the oocytes at least ten times in OR-2 with rapid decanting to remove the immature oocytes and follicle cells.
8. Transfer the oocytes to SW-41 tubes (6 mL per tube).
9. Wash twice with extraction buffer, and fill to 12 mL with extraction buffer.
10. Centrifuged in a SW-41Ti rotor at 38,000 rpm for 1 h at 4°C.
11. The clear supernatant is carefully removed using a 21-gauge needle.

3.6. Extract Fractionation

All procedures are performed at 4°C. Fractions are stored at –70°C.

3.6.1. BioRex 70

1. BioRex 70 column (1 mL packed bed volume per 10 mg extract to be applied) is equilibrated in Buffer B(100).
2. Extract is applied and washed with three column volumes (cv) Buffer B(100).
3. Bound protein is eluted with Buffer B(500).

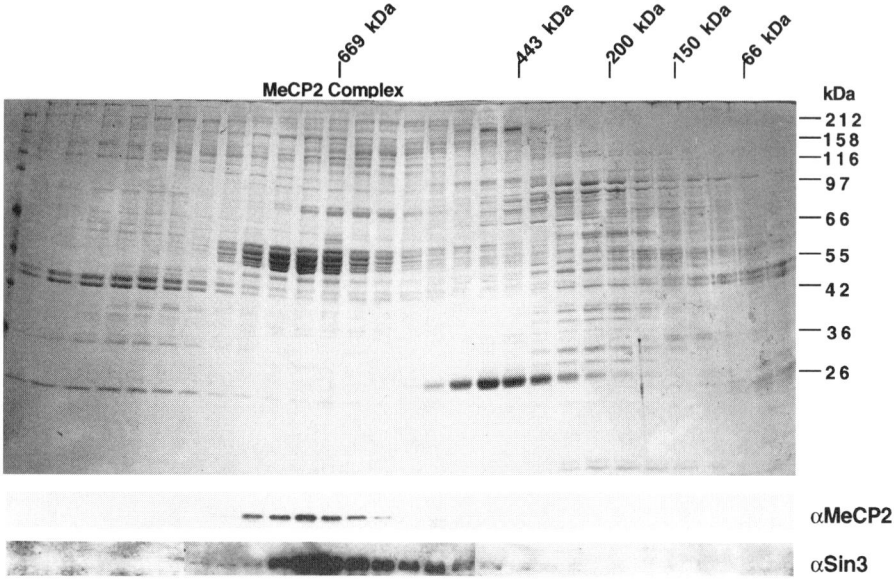

Fig. 3. Cofractionation of MeCP2 and Sin3 by gel filtration. Equal volumes of fractions from the BioRex70 high-salt (0.5 M) step elution separated over a Superose 6 HR 6/30 gel-filtration column were assayed by Western blot for MeCP2 and Sin3 (lower), or stained with Coomassie Blue (upper).

3.6.2. Superose 6 Gel Filtration

1. Equilibrate the Superose 6 HR 10/30 (Pharmacia) FPLC column in Buffer B(500).
2. Filter BioRex 70 B(500) fraction through 0.45-µm syringe filter.
3. Load onto column at 2 mg protein in 500 µL volume.
4. Run FPLC at 0.1 mL/min and collect 250-µL fractions.
5. Assay fractions by Western analysis for MeCP2 and Sin3 (**Fig. 3**).

3.6.3. MonoQ Sepharose

1. Dialyze the BioRex 500 mM elution to <100 mM in 200 vol Buffer B(0).
2. Centrifuge at 12,000g for 20 min to remove insoluble material.
3. Equilibrate a MonoQ sepharose (Pharmacia) HR10/10 (> 20 mg protein to be applied) or HR5/5 (<20 mg protein to be applied) FPLC column in Buffer B(100).
4. Applied the sample to the column, wash with 5 cv Buffer B(100), and eluted in a 20 cv linear gradient from Buffer B(100) to Buffer B(500).
5. Collect fractions of 0.5 cv.
6. Analyze fractions for histone deacetylase activity and by Western and/or Southwestern for MeCP2 and Sin3.

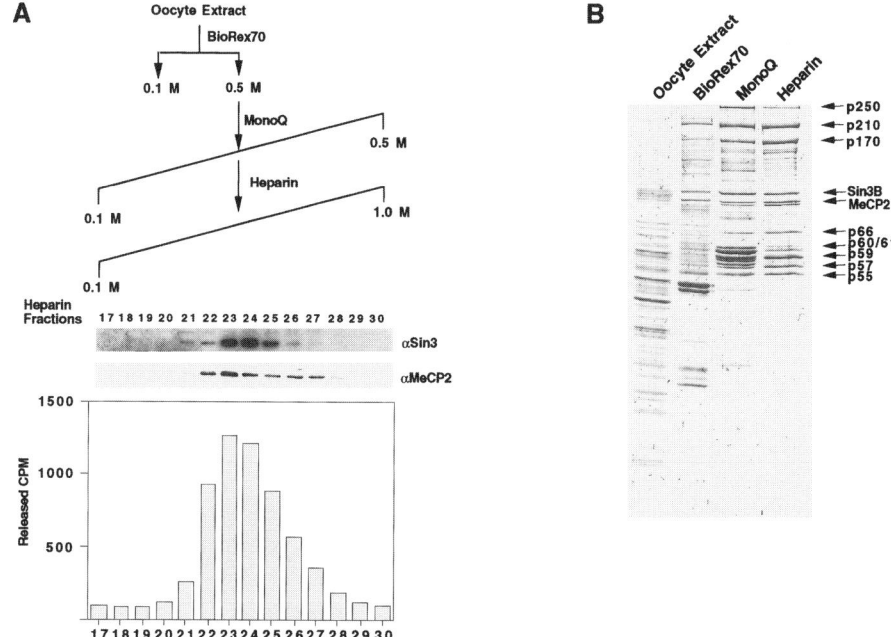

Fig. 4. MeCP2, Sin3, and histone deacetylase copurify. (**A**) Flow chart showing the fractionation of MeCP2 and Sin3 (upper). Fractions were assayed by Western blot for MeCP2 and Sin3 and for histone deacteylase activity. Assays for the final heparin fractionation are shown. (**B**) Coomassie Blue stain of equal amounts of protein (7 ug) from the peak fractions at each step of purification shows that MeCP2 and Sin3 cofractionate with nine additional proteins.

3.6.4. Heparin

The fraction(s) from the MonoQ fractionation containing MeCP2 are used for further purification as follows:

1. Dialyze the sample against 200 volumes Buffer B(0) until the concentration of NaCl is <100 mM.
2. Filter the sample through a 0.45-μm syringe filter.
3. Equilibrate a 1 mL HiTrap Heparin FPLC column (Pharmacia) in Buffer B(100).
4. Load the sample onto the column, wash with 5 cv Buffer B(100), and eluted in a 20 cv linear gradient from Buffer B(100) to Buffer B(1000).
5. Fractions are collected in 0.4 mL volumes and assayed for deacetylase activity and by Western/Southwestern blotting for MeCP2 and Sin3 (**Fig. 4A**).

Using this purification protocol, a complex containing MeCP2, Sin3, and histone deacetylase activity, as well as several currently unidentified proteins (**Fig. 4B**), can be routinely isolated from *Xenopus* oocytes.

4. Notes

1. It is important to remember that for MeCP2, the binding preference is for the symmetrically modified 5-methylcytosine CpG dinucleotide with no other sequence specificity. Therefore, assays must be done in duplicate with probes identical in sequences and differing only in their methylation status, a histone deacetylase assay, and a procedure for the biochemical purification of a MeCP2-histone deacetylase complex.
2. For the purification of the MeCP2-containing deacetylase complex, recombinant yeast Hat1p (generous gift of Dr. M. Parthun, Ohio State University) *(19)* was used to specifically acetylate histone H4. Other specificities can be obtained by using other histone acetyltransferases.
3. The sensitivity of the Southwestern assay depends on how efficiently the protein of interest is able to regain its DNA binding ability after denaturation, immobilization, and renaturation. For MeCP2 from *Xenopus* oocyte extract, it is necessary to precipitate a large protein sample (50 ug) in cold 20% trichloroacetic acid (TCA) followed by a cold acidic acetone wash. For fractions, less sample is needed as the MeCP2 becomes much more concentrated. The sensitivity of the assay for recombinant MeCP2 is about 25 ng.
4. The hybridization can easily be done in a plastic bag on a rotating platform to allow for smaller volumes of hybridization fluid and thus less amount of probe needed.
5. The extent of acetylation can be assayed by Triton acid urea gel *(18)*. This method achieves a specific activity of 2000–5000 dpm per pmol histone H4.
6. We generally use 6 female *X. laevis* to do one oocyte extract preparation yielding around 400 mg soluble extract. In addition, it is very important to thoroughly wash the collagenase-treated oocytes. In addition to the collagenase being a protease itself, it is by no means a pure preparation and may contain additional protease activity. If proteolysis persists despite all efforts at protease inhibition, eggs may be used instead of oocytes following the same extraction and purification scheme presented, starting with loading the eggs into the centrifuge tubes (**Subheading 3.5., step 8**). The reason for using oocytes is that they are easier than eggs to obtain in large numbers, however, eggs do not need the collagenase treatment.

References

1. Kass, S. U., Pruss, D., and Wolffe, A. P. (1997) How does DNA methylation repress transcription? *Trends Genet.* **13,** 444–449.
2. Razin, A. (1998) CpG methylation, chromatin structure and gene silencing: a three-way connection. *EMBO J.* **17,** 4905–4908.
3. Tate, P., Skarnes, W., and Bird, A. P. (1996) The methyl-CpG binding protein MeCP2 is essential for embryonic development in the mouse. *Nature Genet.* **12,** 205–208.

4. Buschhausen, G., Wittig, B., Graessmann, M., and Graessmann, A. (1987) Chromatin structure is required to block transcription of the methylated herpes simplex virus thymidine kinase gene. *Proc. Natl. Acad. Sci. USA* **84,** 1177–1181.
5. Kass, S. U., Landsberger, N., and Wolffe, A.P. (1997) DNA methylation directs a time-dependent repression of transcription initiation. *Curr Biol.* **7,** 157–165.
6. Meehan, R. R., Lewis, J. D., McKay, S., Kleiner, E. L., and Bird, A. P. (1989) Identification of a mammalian protein that binds specifically to DNA containing methylated CpGs. *Cell* **58,** 499–507.
7. Lewis, J. D., Meehan, R. R., Henzel, W. J., Maurer-Fogey, I., Jeppesen, P., Klein, F., and Bird, A. P. (1989) Purification, sequence, and cellular localization of a novel chromosomal protein that binds to methylated DNA. *Cell* **69,** 905–914.
8. Nan, X., Campoy, F. J., and Bird, A. P. (1997) MeCP2 is a transcriptional repressor with abundant binding sites in genomic chromatin. *Cell* **88,** 471–481.
9. Jones, P. L., Veenstra, G. J. C., Wade, P. A., Vermaak, D., Kass, S. U., Landsberger, N., et al. (1998) Methylated DNA and MeCP2 recruit histone deacetylase to repress transcription. *Nature Genet.* **19,** 187–190.
10. Nan, X., Ng, H., Johnson, C. A., Laherty, C. D., Turner, B. M., Eisenman, R. N., and Bird, A. P. (1998) Transcriptional repression by the methyl-CpG-binding protein MeCP2 involves a histone deacetylase complex. *Nature* **393,** 386–389.
11. Struhl, K. (1998) Histone acetylation and transcriptional regulatory mechanisms. *Genes Dev.* **12,** 599–606.
12. Grunstein, M. (1997) Histone acetylation in chromatin structure and transcription. *Nature* **389,** 349–352.
13. Wade, P. A., Jones, P. L., Vermaak, D., and Wolffe, A. P. (1998) A multiple subunit Mi-2 histone deacetylase from Xenopus laevis cofractionates with an associated Snf2 superfamily ATPase. *Curr. Biol.* **8,** 843–846.
14. Heinzel, T., Lavinsky, R. M., Mullen, T., et al. (1997) A complex containing N-CoR, mSin3 and histone deacetylase mediates transcriptional repression. *Nature* **387,** 43–48.
15. Miskimins, W. K., Roberts, M. P., McClelland, A., and Ruddle, F. H. (1985) Use of a protein-blotting procedure and a specific DNA probe to identify nuclear proteins that recognize the promoter region of the transferrin receptor gene. *Proc. Natl. Acad. Sci. USA* **32,** 6741–6744.
16. Vinson, C., LaMarco, K., Johnson, P., Landschulz, W., and McKnight, S. (1988) In situ detection of sequence-specific DNA binding activity specified by a recombinant bacteriophage. *Genes Dev.* **2,** 801–806.
17. Shimamura, A. and Worcel, A. (1989) The assembly of regularly spaced nucleosomes in the Xenopus oocyte S-150 extract is accompanied by deacetylation of histone H4. *J. Biol. Chem.* **264,** 14,524–14,530.
18. Chandler, S. P. and Wolffe, A. P. (1998) Analysis of linker histone binding to mono- and dinucleosomes, in *Methods in Molecular Biology: Chromatin Protocols.* (Becker, P., ed.), Humana Press, Totowa, NJ, pp. 103–112.
19. Parthun, M. R., Widom, J., and Gottschling, D. E. (1996) The major cytoplasmic histone acetyltransferase in yeast: links to chromatin replication and histone metabolism. *Cell* **87,** 85–94.

12

DNA-Methylation Analysis by the Bisulfite-Assisted Genomic Sequencing Method

Petra Hajkova, Osman El-Maarri, Sabine Engemann, Joachim Oswald, Alexander Olek, and Jörn Walter

1. Introduction

The postreplicative methylation of DNA at the C5 position of cytosines is found in a broad spectrum of organisms ranging from prokaryotes to human *(1)*. In prokaryotes the major role of cytosine C5 methylation (like adenine N6 and cytosine N4 methylation) is to protect the genome against DNA degrading nucleases (restriction/modification), whereas in many eukaryotes cytosine C5 methylation (found within CpG dinucleotides) plays a pivotal role in the control of gene expression, inactivation of repetitive sequences, stability of chromosomes, and in cell transformation leading to development of cancer. The growing evidence that the cytosine methylation is also crucial in embryonic development of mammals regulating genomic imprinting, X inactivation and cell differentiation *(2)* has caused a demand for effective methods that would detect this modification with high sensitivity and reliability.

Original methods to detect sequence specific genomic methylation were based on the digestion of DNA by methylation-sensitive restriction enzymes and subsequent Southern-blot hybridization *(3)*. Despite rather high amounts of DNA needed for such experiments and the possibility to investigate just the limited numbers of CpGs situated within suitable restriction sites, the method is still useful as the first indication of methylation in a specific region. To improve the sensitivity, the method was combined with polymerase chain reaction (PCR) amplification *(4)* and subsequent quantification of PCR products *(5,6)*. Although the use of PCR decreased the amount of template

DNA necessary for the analysis, the whole procedure is highly demanding in terms of strictly standardized conditions of DNA preparation and PCR, since quantification is only possible within the exponential phase of amplification. Additionally, incomplete digestion of chromosomal DNA might be a frequent source of artefacts. Another disadvantage of such methods is that they provide data only about the average level of methylation; it is neither possible to discriminate between mosaic or even methylation patterns nor to address hemi-methylation, which remains in general undetected.

The first information about the methylation of cytosine residues irrespective of their sequence context was obtained using a genomic sequencing protocol *(7)*. This method identifies a position of 5-methylcytosine (5-MeC) in the genomic DNA as a site that is not cleaved by any of the Maxam and Gilbert sequencing reactions *(8)* and thus appears as a gap in a sequencing ladder. Although a detailed distribution of methylation in a given sequence can be analyzed by this method, it still requires relatively large amounts of genomic DNA and a certain level of experience in interpreting the sequencing results as bands of varying intensity and shadow bands may occur. An elegant combination of the chemical cleavage method with ligation mediated PCR *(9)* increases the sensitivity, but this modification makes the whole procedure rather laborious and technically challenging.

With a bisulphite genomic-sequencing method *(10,11)*, a qualitatively and quantitatively new approach to methylation analysis has appeared. The bisulphite reaction leads to the conversion of cytosines into uracil residues, which are recognized as thymines in subsequent PCR amplification and sequencing, whereas the modified cytosines do not react and are therefore detected as cytosines. Thus the method allows direct and positive determination of methylation sites in the genomic DNA, as only methylated cytosines are detected as cytosines. Products of PCR-amplified bisulphite-treated DNA can be used directly for sequencing (detection of average methylation status) or cloned and sequenced individually, when the information about the methylation pattern of single molecules is desired. Not only the methylation status of each single molecule but also the pattern of each DNA strand can be investigated, as the strands are no longer complementary following the bisulphite treatment and are amplified and sequenced separately.

Several modifications of the original bisulphite sequencing protocol improving the sensitivity and quality of the results have been published *(11–15)*. In some cases a direct sequence analysis of the PCR products obtained may be desirable to estimate the average methylation at specific sites. For such direct quantitation Gonzalgo and Jones *(16)*, proposed an elegant and simple procedure (Ms-SNuPE). A more sophisticated protocol for direct quantitation of sequencing results is described by Paul et al. *(17)*.

The attributes of high sensitivity, the ability to detect single-molecule methylation patterns as well as the possibility of addressing nonsymmetrical methylation make bisulphite-based genomic sequencing the method of choice for a variety of applications. The following protocol used routinely in our laboratory is based on the previously published procedure *(14)*; several modifications are included leading to easier handling and less time-consuming experimental procedure.

2. Materials
2.1. Embedding of Material into Agarose and Bisulphite Reaction

1. Trypsin: 0.25% [w/v] in PBS (Biochrom).
2. Mineral oil: heavy white mineral oil (Sigma).
3. LMP agarose (SeaPlaque Agarose, FMC).
4. Proteinase K (Boehringer Mannheim).
5. Hydroquinone (Sigma).
6. Phenylmethylsulphonyl fluoride (PMSF) (Sigma).
7. Sodium disulphite: (Sodium metabisulphite) (Merck).

Note: Batches of commercially available sodium bisulphite are mixtures of sodium bisulphite and sodium metabisulphite. The ratio between the substances may vary among different batches. We recommend working with pure sodium metabisulphite, which facilitates accurate preparation of solutions with the desired molarity.

Common laboratory solutions and buffers like sodium dodecyl sulfate (SDS), ethylenediaminetetraacetic acid (EDTA), phosphate-buffered saline (PBS), Tris-HCl, pH8.0, NaOH, TE were prepared according to **ref. 17a**.

2.2. PCR, Purification, and Cloning of PCR Product

1. Taq polymerase (Boehringer Mannheim).
2. Geneclean II (Bio 101), or equivalent method, for purification of PCR fragments from agarose gels.
3. TA cloning kit (Invitrogene) with INVα F` ultra-competent *Escherichia coli* cells.

For some bisulphite–PCR fragments we observed a clonal selection against fully converted templates. In those cases we were able to overcome the problem using a different cloning vector system (pGEM-T, Promega) in combination with competent Sure *E.coli* cells.

3. Method
3.1. Chemistry of the Bisulphite Reaction

The reaction of cytosine residues with sodium bisulphite leading to selective conversion to uracils was first published in early 1970s *(18,19)* for conformational studies of single- and double-stranded regions in DNA and

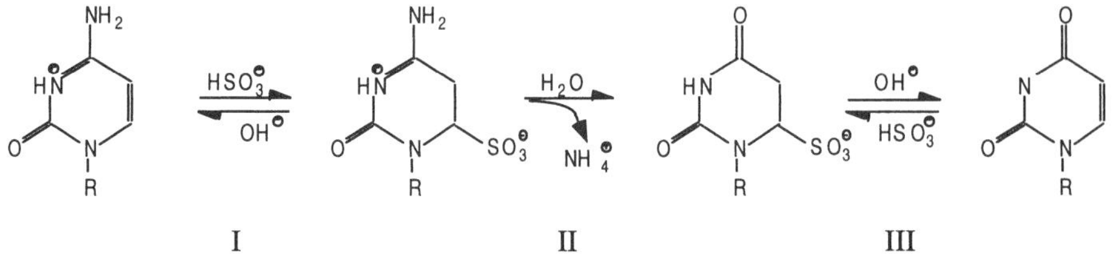

Fig. 1. Chemistry of the reaction steps: I) sulphonation at the position C6 of cytosine, II) irreversible hydrolytic deamination at the position C4 generating 6-sulphonate-uracil, and III) subsequent desulphonation under alkaline conditions. Note that methylation at the position C5 impedes sulphonation at the C6 position (step 1).

RNA *(20–22)*. The reaction generally consists of three major steps (**Fig. 1**): 1) sulphonation, 2) deamination, and 3) desulphonation *(18)*.

1. Reversible sulphonation of cytosine residues to cytosine-6-sulphonate. The sulphonation is favored at low pH and low temperature; at 0°C the equilibrium state is reached within 20 min.
2. Irreversible hydrolytic deamination of cytosin-6-sulphonate to uracil-6-sulphonate. This reaction is favored at higher concentrations of sodium bisulphite and at higher temperatures; the pH optimum is between pH 5.0 and 6.0.
3. Reversible desulphonation of uracil-6-sulphonate to uracil. The elimination reaction is favored at high pH.

Only cytosines in single-stranded DNA (or its components) can be efficiently modified by sodium bisulphite; cytosines in nondenatured, double-stranded DNA are almost refractory to react. Furthermore, under the conditions described, the reaction is highly selective for nonmethylated cytosine residues, which are quantitatively modified (converted to uracils), whereas only 2–3% of 5-methylcytosine residues do react and are converted to thymines *(23)*.

3.2. Principles of the Method

The following protocol is a modified version of the original bisulphite-based methylation-analysis technique described by Frommer et al. *(10)*. As a main difference we routinely embed the material under investigation (i.e., either isolated DNA or intact cells) into low melting point (LMP) agarose. All the following modification steps are performed in the agarose in which the DNA is physically captured. The described modification greatly facilitates the handling of the probes and reduces the loss of DNA in the procedure, thus allowing to work with minute DNA or cell/tissue quantities. The embedding also reduces the reannealing of denatured DNA strands therefore ensuring highest quality and reproducibility in the bisulphite reaction. The principles described in this paragraph refer to a procedure when working with intact cells (*see* **Subheadings 3.3.** and **3.4.**). A modified version of this protocol (*see* **Subheading 3.4.**) should be applied when using purified DNA.

When intact cells are used as a starting material they are embedded into an agarose, lysed, and treated with Proteinase K to make the genomic DNA accessible for subsequent bisulphite treatment. We recommend including an endonuclease restriction step (with a methylation-insensitive enzyme) to obtain smaller DNA fragments (of about 3–6 kb), which enhances the spatial separation of complementary DNA strands after the following denaturation. Agarose-embedded (and -digested) DNA is denatured by alkaline treatment and boiling. Nonmethylated cytosine residues in single strands are subsequently modified in the presence of 2.5 M sodium bisulphite (see experimental procedures concerning preparation of solutions), converted to uracil residues by following alkaline treatment, washed extensively and stored in minimal volumes of TE (**Fig. 1**). The sequence of interest is finally amplified by (nested) PCR. Due to the bisulphite treatment, DNA strands are no longer complementary, and therefore are amplified and analyzed separately (**Fig. 2**). The PCR products can be used directly for sequencing, which allows the quantitation of average values of cytosine methylation at individual positions. Alternatively, the PCR product may be cloned and individual clones sequenced, the latter revealing information about individual chromosomes.

3.3. Preparation of Cells for Bisulphite Treatment

The following procedure should be used when working with limited amount of tissue (containing less than 1 ng of DNA in total) or only a few cells, in which cases the DNA isolation is difficult. When larger quantities of cellular material are available like biopsies, paraffin-embedded tissues, sperm samples, and so on; we recommend first isolating DNA and then proceeding with procedures described in **Subheading 3.4.**

1. When starting with tissue samples, this material should be trypsinized to obtain a single-cell suspension. In cases of individually collected cells (e.g., oocytes, zygotes, etc.) proceed directly to **step 2**.
2. Wash and recover the cells in 1X PBS solution at a maximum density of 60 cells/µL. (In case of oocytes/zygotes, 30–50 cells should be used per agarose bead.)
3. Mix 3 µL of cell suspension with 6 µL of hot (80°C) 2% (w/v) LMP agarose (SeaPlaque Agarose, FMC) (prepared in 1X PBS).
4. Add 500 µL of heavy mineral oil, incubate in boiling water bath for 30 min, and transfer to ice (additional 30 min) to solidify the agarose/cell mixture.
5. Incubate the agarose bead in 500 µL of the lysis solution (10 mM Tris-HCl, 10 mM EDTA, 1% SDS, 20 µg/mL proteinase K) under the mineral oil at 37°C overnight.

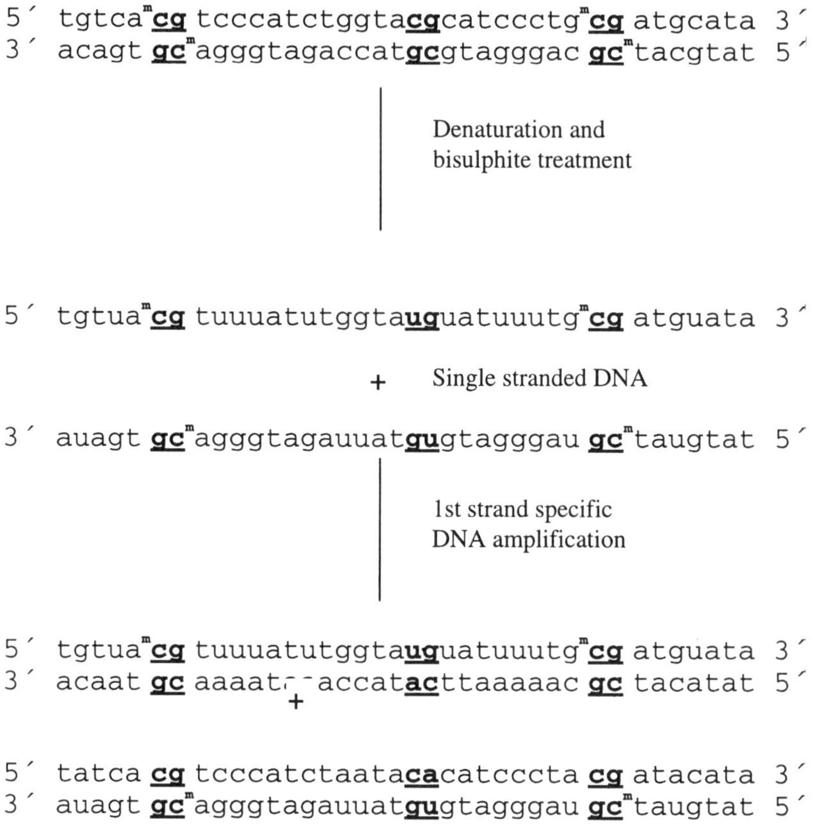

```
5´ tgtca^mcg tcccatctggtacgcatccctg^mcg atgcata 3´
3´ acagt gc^magggtagaccatgcgtagggac gc^mtacgtat 5´
```

Denaturation and bisulphite treatment

```
5´ tgtua^mcg tuuuatutggtauguatuuutg^mcg atguata 3´
```
+ Single stranded DNA
```
3´ auagt gc^magggtagauuatgugtagggau gc^mtaugtat 5´
```

1st strand specific DNA amplification

```
5´ tgtua^mcg tuuuatutggtauguatuuutg^mcg atguata 3´
3´ acaat gc  aaaat--accatacttaaaaac gc tacatat 5´
                +
5´ tatca cg tcccatctaatacacatcccta cg atacata 3´
3´ auagt gc^magggtagauuatgugtagggau gc^mtaugtat 5´
```

Fig. 2. Schematic diagram of the bisulphite conversion of a DNA sequence; note that the upper and lower strands are no longer complementary after the bisulphite treatment (a, adenine; c, cytosine; ^{m}c, 5-methyl-cytosine; g, guanine; t, thymine; u, uracil).

6. Remove the lysis solution and the oil and inactivate proteinase K in 500 μL of TE, pH 7.0, containing 40 μg/mL PMSF 2 × 45 min at room temperature (RT). (This step is optional.)
7. Remove the solution and wash with 1 mL of 1X TE, pH 9.0, 2 × 15 min (i.e. 2 washes, 15 min each).
8. Equilibrate against 100 μL of restriction buffer 2 × 15 min.
9. Remove the solution and add 100 μL of 1X restriction buffer containing 20 units of restriction endonuclease and incubate overnight. (Alternatively add 50 units for 1 h digestion.)
10. Remove the restriction buffer and incubate with 500 μL of 0.4 M NaOH 2 × 15 min.
11. Wash with 1 mL of 0.1 M NaOH for 5 min.

12. Remove all the solution and overlay with 500 µL of mineral oil.
13. Boil the beads in a water bath for 20–30 min to separate the individual DNA strands.
14. Chill on ice for 25 min to re-solidify the agarose bead.
15. Prepare the bisulphite/hydroquinone solution according to **Subheading 3.4., step 6**.
16. Add 500 µL of the (ice-cold) bisulphite/hydroquinone solution. The agarose bead should be in the (lower) aqueous phase.
17. Proceed according to **Subheading 3.4., step 9**.

3.4. Bisulphite Treatment of Isolated DNA

Bisulphite and hydroquinone solutions are light-sensitive, thus should be protected from light in all steps.

1. Digest genomic DNA with a suitable restriction enzyme (which does not cut within the region to be amplified) in a volume of 21 µL. In order to achieve a complete bisulphite conversion, we recommend using not more than 700 ng DNA for the restriction, so that the DNA content of each (later on) formed agarose-DNA bead does not exceed 100 ng, *see* **step 8**.
2. Boil for 5 min in a water bath.
3. Chill on ice and quickly spin down.
4. Add 4 µL of 2 M NaOH (final concentration 0.3 M NaOH) and incubate 15 min at 50°C.
5. Mix with 2 vol (50 µL) of melted (50–65°C) 2% (w/v) LMP agarose (SeaPlaque Agarose, FMC; prepared in water).
6. Prepare 2.5 M bisulphite solution, pH 5.0, as follows: dissolve 1.9 g of sodium bisulphite in a mix of 2.5 mL H_2O and 750 µL of $2M$ NaOH (freshly prepared), dissolve 55 mg of hydroquinone in 500 µL of H_2O at 50°C, and mix both solutions.
7. Pipet 1 mL of the bisulphite/hydroquinone solution into a 2-mL Eppendorf tube and overlay with 750 µL of heavy mineral oil (tubes should be kept for 30 min on ice before proceeding).
8. Pipet up to seven 10 µL-aliquotes of the DNA-agarose mixture into ice-cold mineral oil to form beads. (Each bead should contain up to 100 ng of DNA.) Make sure that all beads have entered the aqueous phase; beads can be pushed into the bisulphite solution using a pipet tip.
9. Leave on ice for 30 min.
10. Incubate at 50°C for 3.5 h.
11. Remove all solutions; wash with 1 mL of 1X TE, pH 8.0 for 4 × 15 min.
12. Add 500 µL of 0.2 M NaOH 2 × 15 min.
13. Remove NaOH solution and wash with 1 ml of 1X TE, pH 8.0, 3 × 10 min. Store in a small volume of TE, pH 8.0, at 4°C (beads are stable for at least several weeks).
14. Prior to amplification wash the beads with H_2O for 2 × 15 min.

3.5. General Recommendations

3.5.1. Primer Design

The guidelines for primer design for the amplification of bisulphite-treated DNA presented here concern methylation patterns found in mammalian genomes (i.e., methylation mainly at CpG sites). Different requirements for primer design have to be considered when analyzing methylation patterns in organisms with a broader methylation spectrum (as CpNpG or nonsymmetrical cytosine methylation in plants or fungi, and so on; *see* **Subheading 4.**).

1. A bisulphite-treated DNA sequence should be generated using any word processor to replace all Cs for Ts except at CpG sites (e.g., for DNA methylation patterns in mammalian genomes). Any primer designing software that will help to avoid any hairpin structures and possible primer dimers can use this modified sequence.
2. The length of the oligos should be at least 20 nucleotides and up to 25–30 nucleotides.
3. The primers should be located in an originally cytosine-rich region so that they selectively amplify converted DNA.
4. Overlapping of the primers with CpG dinucleotides should be strictly avoided especially at the 3' end of the oligos.
5. Extensive T and A stretches in both primers, which are typical for bisulphite-treated DNA, should be avoided to minimize the formation of primer dimers.

3.5.2. Optimizing PCR Conditions

1. The PCR conditions for amplifying bisulphite-treated material should be carefully optimized. The bisulphite treatment reduces the sequence specificity (by changing all cytosines to uracils) and thus the selectivity for primer annealing.
2. It is recommended that the length of the product does not exceed 600–700 bp as longer fragments may be more difficult to amplify from a bisulphite-treated DNA (due to depurination of DNA as a result of low pH during the bisulphite treatment).
3. A nested or at least a seminested approach for amplifying the target region is recommended to increase the sensitivity when working with limited numbers of cells and to ensure the specificity of the product.
4. To avoid any contamination with previous PCR products, the bisulphite treatment and handling of the DNA or cells should be carried in a separate room and using separate pipets.
5. A gradient PCR cycler is recommended to optimize the annealing temperature.

3.5.3. Cloning and Sequencing

1. To increase the efficiency of cloning, the specific PCR product should be purified from any unspecific band(s) or primer dimers by agarose-gel elution.

Table 1
Frequently Encountered Problems

Problem	Recommended solution
1. The hydroquinone solution turns red.	a. Protect the solution from light and do not heat over 50°C to dissolve.
2. During incubation of the bisulphite-hydroquinone solution on ice crystals appear.	a. This will not affect the results: proceed normally.
3. The beads disappear after entering the bisulphite solution.	a. The mineral oil layer is not cold enough; the tubes containing the mineral oil should be pre-incubated on ice for a longer period (at least 20 min) or alternatively they can be kept at –20°C for 10 min, then the bisulphite solution added after the formation of the beads. b. Use only heavy mineral oil. c. Increase the concentration of LMP agarose.
4. No PCR product is obtained.	a. Inefficient bisulphite conversion (*see* **Subheading 3.4.**). b. The amount of template is not sufficient. c. The PCR product is too long for bisulphite-treated DNA; design primers that amplify smaller fragments. d. Use nested primers to increase the sensitivity and the yield of the amplification. e. Try different set of primers.
5. The PCR product is difficult to clone.	a. If a T/A cloning vectors are used: prior to ligation, incubate the PCR product with additional amount of Taq polymerase and dATP (this would help to add a flanking A to the 3′ end of the product). b. Try different cloning vectors and *E. coli* strains.
6. Unconverted sequences are frequently observed.	a. Primers are not selective enough for converted DNA. The primers should be located in a C-rich region to increase the specificity of amplification for converted sequences. b. Incomplete bisulphite conversion may be caused by excess of DNA in reaction; the maximum recommended amount of DNA per bead is 100 ng. c. DNA was not properly denatured, make sure that denaturation steps and desulphonation steps are properly made using fresh NaOH solution. d. The bisulphite and hydroquinone solutions should be prepared fresh and stored no longer than 24 h before use.

2. Cloning of the PCR product can be improved by additional incubation of the purified product in presence of dATP and Taq polymerase for 5 min at 95°C followed by 60 min at 72°C. This will increase the percentage of DNA molecules with flanking As at the 3′ end.
3. To verify the presence of the correct insert in plasmids, we routinely use a colony PCR protocol. Products of the correct size can be directly sequenced using internal primers.
4. According to our experience, blue/white screening of colonies is not always reliable (especially when short fragments are cloned). In such situation, we recommend analyzing all colonies, as the blue ones may contain an insert.

3.6. Drawbacks of the Bisulphite-Based Methylation Analysis

Although the bisulphite-based methylation analysis is a powerful tool to obtain detailed genomic-methylation data, it is connected with specific experimental or technical problems, which are briefly discussed here.

1. In order to perform a bisulphite-based methylation analysis, a detailed sequence information of the genomic region of interest is required.
2. The upper and lower strands of the bisulphite-treated DNA samples are analyzed separately. Therefore it is impossible (except for a single-cell approach) to obtain data about the original double-stranded DNA.
3. Amplifications (or cloning of PCR products) from the upper and lower strand may not work equally well in each case.
4. In case of analyzing methylation patterns of unknown distribution as, e.g., in plants and fungi, it may be difficult to design primers for the PCR amplification of the bisulphite-treated DNA. In this case primers can be designed that contain either C or T at the respective positions. However, the use of such primers with "wobble" positions greatly reduces the sensitivity of the PCR reaction and may cause a bias in the amplification towards specific (mostly not fully converted) products.
5. A systematic analysis *(24)* nicely demonstrated that the choice of primers might cause a bias in the PCR reaction, such that either the low or highly methylated template DNA is predominantly amplified. Another selection against a specific subset of PCR products may occur during the cloning procedure. The problem of biased amplification or cloning has to be tested individually and several control experiments should be carried out. First, different templates with a known content of methylated cytosine residues should be mixed in different ratios and the bisulphite treatment and amplification steps should be carried out as usual. The distribution of nonconverted and converted cytosine residues in the analyzed products will then allow determination of whether and in which magnitude a bias occurs. Furthermore, independent experiments (including different techniques) should be used to analyze the methylation state of a given template, e.g., conventional Southern-blot hybridization or the Ms-SNuPE assay *(16)*. Both experiments may be very helpful to obtain an independent impression of the real ratio of modified and unmodified cytosines within the sequence of interest.

Acknowledgments

This work was supported by the Deutsche Forschungsgemeinschaft Wa1029/1 and the European Union BMH4-CT96-0050.

References

1. Jost, J. P. and Saluz, H. P., eds. (1993) *DNA Methylation: Molecular Biology and Biological Significance.* Birkhauser Verlag, Basel, Switzerland.
2. Li, E. B. C. and Jaenisch, R. (1992) Targeted mutation of the DNA methyltransferase gene results in embryonic lethality. *Cell* **69,** 915–926.
3. Southern, E. M. (1975) Detection of specific sequences among DNA-fragments separated by gel electrophoresis. *J. Mol. Biol.* **98,** 503–517.
4. Singer, S. J., Robinson, M. D., Bellve, A. R., Simon, M. I., and Riggs, A. D. (1990) Measurement by quantitative PCR of changes in HPRT, PGK-1, PGK-2, APRT, MTase, and Zfy gene transcripts during mouse spermatogenesis. *Nucleic Acids Res.* **18,** 1255–1259.
5. Kafri, T., Ariel, M., Brandeis, M., Shemer, R., Urven, L., McCarrey, J., et al. (1992) Developmental pattern of gene specific DNA methylation of the mouse embryo and germ line. *Genes Dev.* **6,** 705–714.
6. Brandeis, M., Kafri, T., Ariel, M., Chaillet, J. R., McCarrey, J., Razin, A. and Cedar, H. (1993) The ontogeny of allele-specific methylation associated with imprinted genes in the mouse. *EMBO J.* **12,** 3669–3677.
7. Maxam, A. M. and Gilbert, W. (1980) Sequencing end-labeled DNA with base-specific chemical cleavages. *Methods Enzymol.* **65,** 499–560.
8. Church, G. M. and Gilbert, W. (1984) Genomic sequencing. *Proc. Natl. Acad. Sci. USA* **81,** 1991–1995.
9. Pfeifer, G. P., Steigerwald, S. D., Mueller, P. R., Wold, B. and Riggs, A. D. (1989) Genomic sequencing and methylation analysis by ligation mediated PCR. *Science* **246,** 810–813.
10. Frommer, M., McDonald, L. E., Millar, D. S., Collis, C. M., Watt, F., Grigg, G. W., et al. (1992) A genomic sequencing protocol that yields a positive display of 5-methylcytosine residues in individual DNA strands. *Proc. Natl. Acad. Sci. USA* **89,** 1827–1831.
11. Clark, S. J., Harrison, J., Paul, C. L., and Frommer, M. (1994) High sensitivity mapping of methylated cytosines. *Nucleic Acids Res.* **22,** 2990–2997.
12. Feil, R., Walter, J., Allen, N.D., and Kelsey, G. (1994) Developmental control of allelic methylation in the imprinted mouse *Igf2 and H19* genes. *Development* **120,** 2933–2943.
13. Raizis, A. M., Schmitt, F., and Jost, J. P. (1994) A bisulphite method of 5-methyl-cytosine mapping that minimises template degradation. *Anal. Biochem.* **226,** 161–166.
14. Olek, A., Oswald, J., and Walter, J. (1996) A modified and improved method for bisulphite based cytosine methylation analysis. *Nucleic Acids Res.* **24,** 5064–5066.

15. Paulin, R., Grigg, G. W., Davey, M. W., and Piper, A. A. (1998) Urea improves efficiency of bisulphite-mediated sequencing of 5<<-methylcytosine in genomic DNA. *Nucleic Acids Res.* **26,** 5009–5010.
16. Gonzalgo, M. and Jones, P. A. (1997) Rapid quantitation of methylation differences at specific sites using methylation-sensitive single nucleotide primer extension (Ms-SNuPE). *Nucleic Acids Res.* **25,** 2529–2531.
17. Paul, C. L. and Clark, S. J. (1996) Cytosine methylation: quantitation by automated genomic sequencing and GENESCAN analysis. *BioTechniques* **21,** 126–133.
17a. Sambrook, G., Fritsch, E. F., and Maniatis, T. (1988) *Molecular Cloning: A Laboratory Manual*, 2nd ed. Cold Spring Harbor Laboratory Press, Cold Spring Harbor, New York.
18. Hayatsu, H., Wataya, Y., Kai, K., and Iida, S. (1970) Reaction of sodium bisulphite with uracil, cytosine, and their derivatives. *Biochemistry* **9,** 2858–2864.
19. Shapiro, R., Braverman, B., Louis, J. B., and Servis, R. E. (1973) Nucleic acid reactivity and conformation: reaction of cytosine and uracil with sodium bisulfite. *J. Biol. Chem.* **248,** 4060–4064.
20. McLaren, A., Gonos, E. S., Carr, T., and Goddard, J. P. (1993) The conformation of tRNA genes. Chemical modification studies. *FEBS Lett.* **13,** 177–180.
21. Goodchild, J., Fellner, P., and Porter, A. G. (1975) The determination of secondary structure in the polyC tract of encephalomyocarditis virus RNA with sodium bisulphite. *Nucleic Acids Res.* **2,** 797–805.
22. Kelly, J. M., Goddard, J. P., and Maden, E. H. (1978) Evidence on the conformation of HeLa-cell 5.8S ribosomal ribonucleic acid from the reaction of specific cytidine residues with sodium bisulphite. *Biochem. J.* **173,** 521–532.
23. Wang, R. Y.-H., Gehrke, C. W., and Ehrlich, M. (1980) Comparison of bisulfite modification of 5-methyldeoxycytidine and deoxycytidine residues. *Nucleic Acids Res.* **8,** 4777–4790.
24. Warnecke, P. M., Stirzaker, C., Melki, J. R., Millar, D. S., Paul, C. L., and Clark, S. J. (1997) Detection and measurement of PCR bias in quantitative methylation analysis of bisulphite-treated DNA. *Nucleic Acids Res.* **25,** 4422–4426.

13

Measuring DNA Demethylase Activity In Vitro

Moshe Szyf and Sanjoy K. Bhattacharya

1. Introduction
1.1. Alternative Assays for Demethylation

The reaction catalyzing direct demethylation of DNA involves the removal of a methyl group residue from the 5′ position on cytosine; the products of the reaction are nonmethylated cytosine in the dinucleotide CpG and methanol *(1)*. The study of the proteins involved in demethylation requires an assay for measuring enzymatic DNA demethylation. A number of assays were previously described for determining the state of methylation of CpG sequences DNA. For example, certain restriction enzymes such as *Hpa*II or *Hha*I recognize subsets of CpG sequences only when the C is not methylated; thus, cleavage by this enzymes indicates demethylation of their recognition sequences *(2)*. This assay is, however, obviously indirect and can only measure the state of methylation of a subset of CG sequences contained in the enzyme-specific recognition site. An additional problem is that this assay does not differentiate between DNA that is directly demethylated, and repair processes that remove methylated cytosines in DNA and replace them with other unmethylated cytosines found in the extracts. An additional assay is the bisulfite-mapping method, which can determine the state of methylation of cytosines at a single nucleotide resolution *(3)*. This assay is based on the fact that nonmethylated cytosines are modified by bisulfite and converted to thymidine, whereas methylated cytosines are protected. This assay, similar to the restriction enzyme-based assays, is indirect; it does not measure demethylation but rather the conversion of cytosines. This assay can not differentiate as well between true demethylation and repair. Moreover, this is an extremely labor-intensive assay that involves polymerase chain reaction (PCR) amplification and subcloning and is not feasible as a routine assay for measuring enzymatic activity in multiple samples or for assaying a large number of fractions during chromatographic fractionation.

We have developed an assay that measures directly the conversion of 5-methylcytosine to cytosine that is based on previously described nearest-neighbor analysis of CpG methylation *(4)*. To measure directly the conversion of 5-mdCMP in DNA to dCMP, we synthesize a completely methylated ^{32}P-labeled [mdC^{32}pdG]n double-stranded oligomer. Following incubation with the different fractions, the DNA is purified and subjected to cleavage with micrococcal nuclease to 3′ mononucleotides. The 3′ labeled mdCMP and dCMP are separated by thin-layer chromatography (TLC) and the conversion of mdCMP to dCMP is directly determined. Since only cytosines that were originally ^{32}P-labeled are detected, this assay will exclude the possibility that new unmethylated cytosines were introduced by a repair process. This assay provides therefore a stringent test for *bona fide* demethylation. As a control, we synthesize an unmethylated [dC^{32}pdG]n double-stranded oligomer (**Fig. 1**).

The main advantage of the ^{32}P-labeled [mdC^{32}pdG]n double-stranded oligomer in enzymatic analysis is that it is a well-defined simple sequence. However, it does not resemble the normal situation of CpGs in the genome, which are interspersed among complex and varied sequences. In order to address this concern, we synthesized a plasmid based substrate that is fully methylated at its cytosines and bears a ^{32}P-labeled phosphodiester bond between all the Cs and Gs in its sequence *(5)*. Since plasmids bearing different genes are available, they could be used to study true direct demethylation of different genes in vitro. This chapter describes the protocol that is used in our laboratory to prepare these substrates and how they are used to study demethylation in vitro.

2. Materials

1. [^{32}P-α]- dGTP: 3000 Ci/mmol (New England Nuclear BLUSIHH).
2. dNTP 100 mM (Roche cat. no. 1 277 049).
3. d^{methyl}CTP (Roche cat. no. 757 047).
4. Nap5 desalting columns (Pharmacia 17-0853-01).
5. DNA polymerase Klenow fragment (Roche cat. no. 997 455).
6. T4 polynucleotide kinase (2,500 units) and 10X Reaction buffer (New England Biolabs cat. no. 201L).
7. 100 mM ATP (Roche cat. no. 1 140 965).
8. 100 bp DNA ladder (New England Biolabs cat. no. 323-1S).
9. QIAquick PCR purification kit (Quiagen cat. no. 28104).
10. Thin-layer chromatography plate Polygram CEL 4000 0.1 mm microcristalline cellulose 20 × 20 cm Macherey-Nagel, distributed by Alltech (Art.-Nr. 801113).

DNA Demethylase Activity

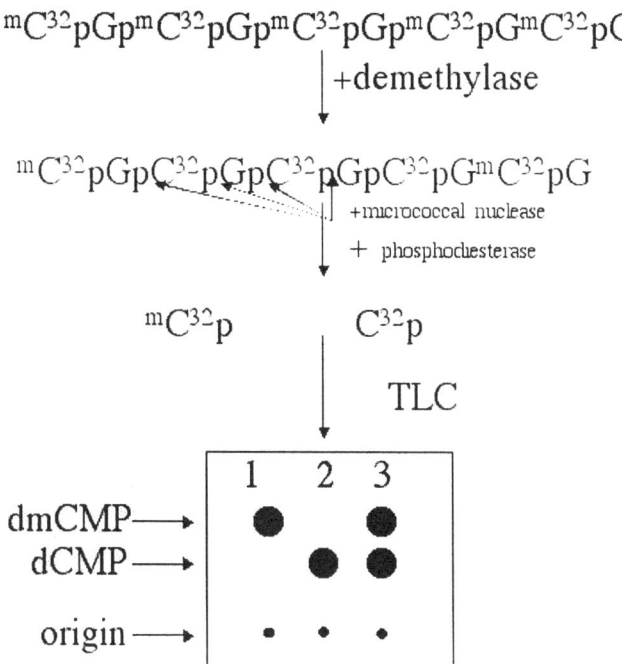

Fig. 1. A scheme of an assay of demethylase activity. A ^{32}P labeled and methylated oligo $mC^{32}pG$ substrate (the substrate is double stranded however to simplify the scheme only one strand is shown) is incubated with a demethylase preparation. Following demethylation, the DNA is digested with 3′ mononucleases micrococcal nuclease and spleen phosphodiesterase. The positions of cleavages are indicated by arrows. The resulting ^{32}P labeled methylated dmCMP and nonmethylated dCMP cytosine mononucleotides are separated by TLC. Lane 1 is a methylated control, lane 2 is an unmethylated control, lane 3 is a partially demethylated sample. The position of the origin (sample loading position), dmCMP and dCMP are indicated by arrows.

11. TLC developing chamber for plates up to 20 cm (Fisher cat. no. 05-719-21B and 05-719-21A).
12. Microfuge.
13. Phenol.
14. Chloroform.
15. High prime labeling kit (Roche cat. no. 1-585-584).
16. Yeast tRNA 10 µg/µL (Roche cat. no. 109 495).
17. Ethanol 95%.
18. Ethanol 75%.
19. SssI methylase (New England Biolabs cat. no. 2265).

20. Micrococcal nuclease (resuspend in water to final concentration of 150 U/μL) (Pharmacia cat. no. 70196Y).
21. Spleen phosphodiesterase (Sigma P6897).
22. Isobutiric acid.

2.1. Solutions and Buffers

1. 10X Polymerase buffer: 0.5 M NaCl, 66 mM Tris-HCl, pH 7.4, 66 mM MgCl$_2$, 10 mM DTT.
2. 10X Micrococcal nuclease buffer: 250 mM Tris-HCl, 10 mM CaCl$_2$.
3. Separation solvent for chromatography: isobutyric acid: NH$_3$: H$_2$O (132:2:40) solvent.
4. Buffer L: 10 mM Tris-HCl, pH 7.5, 10 mM MgCl$_2$.
5. Buffer L: 50 mM NaCl.
6. Tris-EDTA buffer: 10 mM Tris-HCl, pH 7.6, 1 mM EDTA.

3. Methods

3.1. Synthesis of ^{32}P-labeled [mdC^{32}pdG]n and [dC^{32}pdG]n Double-Stranded Oligomers

1. The [mdCpdG]$_{20}$ and [mdCpdG]$_6$ oligonucleotides and the [dCpdG]$_{20}$ and [dCpdG]$_6$ unmethylated control oligonucleotides are synthesized by commercial oligonucleotide suppliers. To generate a double-stranded oligonucleotide, we denature 100 μg of oligonucleotide in buffer L (50 mM NaCl) at 100°C for 10 min and allow the DNA to renature by allowing the solution to cool down slowly for 2 h.
2. A 100–300 bp mdCpdG or dCpdG repeat is generated by ligating either the [mCpG]$_{20}$ or the [dCpdG]$_{20}$ oligonucleotide and removing the nonligated primers using a QIAquick PCR purification kit.
3. Ligation requires a 5′ phosphate group. Oligonucleotides supplied by most suppliers do not have a 5′ phosphate. The 5′ phosphate could be added using a kinase enzyme and ATP. Incubate either 10 μg of [mdCpdG]$_{20}$ or [dCpdG]$_{20}$ oligonucleotide in a 50 μL reaction mixture containing 5 μL (50U) T4 polynucleotide kinase, 5 μL of 10X reaction buffer supplied by the manufacturer, and 0.5 μL of 100 mM ATP for 1 h at 37°C.
4. Ligation is carried out in the same buffer. Add 500 U of T4 DNA ligase and incubate the reaction at 16°C for 1 h. To verify the success of the ligation reaction fractionate a 1 μL sample on a 1.5% agarose gel alongside a 100 bp DNA ladder. If the average size of the ligated product is at least 100 bp, proceed to clean up the DNA from nonligated oligonucleotides using a QIAquick Spin column as recommended by the manufacturer.
5. To label either the mdCpdG or the control dCpdG polymers with radioactive ^{32}P, incubate 1 μg of either the ligated mdCpdG (for the methylated substrate) or dCpdG polymers (unmethylated substrate), 0.2 μg of either [mdCpdG]$_6$ (for methylated substrtate) or [dCpdG]$_6$ oligonucleotide (for unmethylated substrate),

and 9 µL of water for 10 min at 100°C. Place the reaction on ice and add 2 µL of 10X DNA polymerase buffer, 2 µL (20 units) of DNA polymerase I Klenow fragment, 1 µL of mdCTP (20 mM) for the methylated substrate or 1 µL of dCTP (20 mM) for the unmethylated substrate, and 2.5 µL of [^{32}P-α]dGTP and 2.5 µL of water. Incubate for 1 h at 37°C. The reaction is terminated by adding 500 µL of Tris-EDTA buffer 250 µL of phenol and 250 µL of chloroform. Vortex and spin at maximum speed in a microfuge. Transfer the aqueous phase into a new tube, avoiding transfer of any of the organic phase.
6. The labeled methylated and nonmethylated substrates are separated from the nucleotides using a Nap5 column. Load the 500 µL reaction mixture on the column. Elute the methylated DNA by applying a sequence of six 0.5-mL aliquots of buffer L and collect each 0.5 mL fraction in a separate microfuge tube. To determine which fraction contains the methylated DNA, transfer a 1-µL sample of each of the fractions into a tube containing 2 mL of 10% TCA and determine the number of counts incorporated into DNA in each fraction. Under these conditions the DNA elutes at the second and third fractions.
7. Combine the fractions and store the labeled DNA at −20°C for future demethylase assays. The concentration of the DNA is 1 ng/µL and its specific activity is typically $1-2 \times 10^7$ dpm/µg.

3.2. Synthesis of ^{32}P Labeled [mdC^{32}pdG] and [dC^{32}pdG] Plasmids

1. To label 1 µg of a plasmid with ^{32}P at the 5′ phosphate of dG, use a standard random-primed DNA-labeling kit and [^{32}P-α]-dGTP as the radioactive radionucleotide as recommended by the manufacturer. We use the kit provided by Boehringer. Prepare two labeled plasmid preparation. One preparation will be used as an unmethylated control.
2. Following incubation for 1 h at 37°C add 150 µL of Tris-EDTA, 10 µg of yeast tRNA, 15 µL of 5 M NaCl, 75 µL of phenol, and 75 µL of chloroform. Vortex and spin at max speed in a microfuge. Transfer the aqueous phase to a fresh tube and add 300 µL of ethanol. Incubate the samples in a dry ice/ethanol bath for 15 min. Spin the samples in a microfuge at maximum speed for 15 min. Discard the supernatant, taking care not to disrupt the pellet, add 300 µL of 70% ethanol, and spin at maximum speed for 15 min. Discard the supernatant and dry the tube using a Kimwipe. Resuspend the pellet in 70 µL water.
2. To methylate the plasmid, add 5 µL of SssI methylase, 10 µL of the buffer supplied by the manufacturer (NEB2), 1 µL of 3.2 mM AdoMet (supplied by the manufacturer). Incubate at 37°C for 3 h.
3. It is essential to achieve full methylation of the substrate. In our experience, three rounds of methylation are required. Perform a phenol-chloroform extraction and ethanol-precipitation step between rounds of methylation.
4. Following the third round of methylation, add 150 µL water, 250 µL phenol, and 250 µL chloroform to the reaction mixture and continue to clean the DNA on a Nap5 column as described in **Subheading 3.1., steps 6** and **7**.

5. To prepare the nonmethylated control, incubate the radioactive plasmid under the same conditions but in the absence of the methylase enzyme. Follow all the steps described above.

3.3. Demethylation Assay

1. For each reaction, add 20 µL buffer L, 2.5 µL of either [mdC^{32}pdG] oligonucleotide or the [mdC^{32}pdG]-plasmid DNA (1 ng/µL), and 20 µL of either buffer L for the control sample or 20 µL of each of the different test samples into each tube. As an additional control, incubate 2.5 µL of either nonmethylated dCpG oligomer or the nonmethylated plasmid with 40 µL buffer L. Allow the reaction to continue for different time points (30 min up to 24 h).
2. Terminate the reaction by adding 100 µL Tris-EDTA, 10 µL tRNA (10 µg/µL), 10 µL NaCl 5 M, 50 µL phenol, and 50 µL chloroform. Vortex, spin at maximum speed in a microfuge, transfer the aqueous phase, and ethanol-precipitate as described in **Subheading 3.2., step 2**.
3. Following ethanol precipitation, resuspend each pellet in 8 µL water. Ad 1 µL 10X micrococcal nuclease buffer and 1 µL micrococcal nuclease. Incubate for 12 h at 37°C.
4. Add 1 µL of Spleen phosphodiesterase and incubate for additional 2 h.
5. Load 3 µL of the samples on a straight line 2 cm from the bottom of a 20 cm × 20 cm TLC plates, spacing them at 1.5-cm intervals. Air-dry the spots.
6. The chromatography solvent should be equilibrated in the chromatography tank 24 h before use. The same solvent could be used for 2 mo. Add 132 mL isobutyric acid and 40 mL water to the chromatography chamber, close the lid almost completely, and add using a pipettor 2 mL of NH$_3$. Seal the lid tightly using vacuum grease and allow the solvent to equilibrate over night.
7. Place the TLC plate in the tank. Seal the lid tightly and allow the chromatography to develop for 6 h or till the solvent reaches 0.5 cm from the top of the plate. Dry the plate and expose to an X-ray film or a phosphorimager plate.
8. Once exposed, the positions of methyldCMP and dCMP could be easily distinguished with reference to the control lanes as shown in **Fig. 1**. The relative ratio of the methylated and nomethylated cytosines is quantified by standard densitometry.

Acknowledgments

The research reported here was supported by grants from the Medical Research Council of Canada and the National Cancer Institute of Canada to MS.

References

1. Ramchandani, S., Bhattacharya, S. K., Cervoni, N., and Szyf, M. (1999) DNA methylation is a reversible biological signal. *Proc. Natl. Acad. Sci. USA* **96**, 6107–6112.

2. Waalwijk, C. and Flavell, R. A. (1978) DNA methylation at a CCGG sequence in the large intron of the rabbit b-globin gene: tissue specific variations. *Nucleic Acids. Res.* **5,** 4631–4641.
3. Clark, S. J., Harrison, C. L., Paul, C. L., and Frommer, M. (1994) High sensitivity mapping of methylated cytosines. *Nucleic Acids Res.* **22,** 2990–2997.
4. Szyf, M. Theberge, J., and Bozovic, V. (1995) Ras induces a general DNA demethylation activity in mouse embryonal P19 cells. *J. Biol. Chem.* **270,** 12,690–12,696.
5. Bhattacharya, S. K., Ramchandani, S., Cervoni, N., and Szyf, M. (1999) A mammalian protein with specific demethylase activity for mCpG DNA. *Nature* **397,** 579–583.

14

Extracting DNA Demethylase Activity from Mammalian Cells

Moshe Szyf and Sanjoy K. Bhattacharya

1. Introduction
1.1. The Demethylase Reaction: What is Demethylase?

Purification of an enzymatic activity requires a simple and relatively expeditious assay of its activity. However, the nature of the demethylase reaction has been elusive for decades. Although a large body of evidence supported the hypothesis that active demethylation takes place during development and differentiation *(1)*, the nature of the reaction was unknown. The main problem with understanding demethylaion of DNA is that true demethylation of DNA would involve cleavage of a stable carbon-carbon bond, which had been considered highly unlikely. Different laboratories have suggested that demethylation of DNA is accomplished by different repair mechanisms. These alternative routes involve either a cleavage of the bond between the methylated cytosine base and the deoxyribose *(2)* or nucleotide excision *(3)* (**Fig. 1**). We have recently shown that mammalian cancer-cell lines bear a *bona fide* demethylase activity and we defined the reactants and products of the demethylation reaction. The demethylation reaction involves the hydrolytic cleavage of the bond between the methyl-carbon and the carbon at the 5 position of the cytosine ring (**Figs. 1** and **2**) producing unmethylated cytosine while the methyl group is released as methanol *(4)*.

1.2. Detection of the Products of the Demethylation Reaction

The products of the demethylation reaction are unmethylated cytosines residing in the dinucleotide sequence CpG and methanol (**Fig. 2**). A number of indirect assays are available for determining the state of methylation of

From: *Methods in Molecular Biology, vol. 200: DNA Methylation Protocols*
Edited by: K. I. Mills and B. H. Ramsahoye © Humana Press Inc., Totowa, NJ

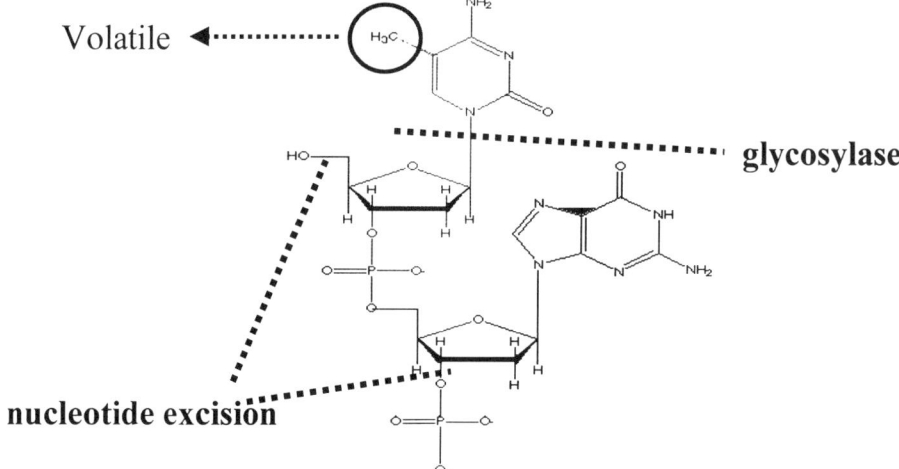

Fig. 1. Possible demethylation mechanisms. The figures shows a methylated CpG dinucleotide. DNA could be demethylated either directly or indirectly. Indirect demethylation might involve either removing the methyl-C base by a glycosylase or removing the methyl-CpG dinucleotide by a nucleotide excision mechanism. The possible cleavage points are indicated. The damaged DNA is repaired by the repair machinery, introducing unmethylated cytosines into the damaged region. Alternatively, the methyl group *per se* could be removed by cleavage of the carbon-carbon bond between the carbon at the 5′ position of cytosine and the methyl group. The methyl group leaves as a volatile residue.

DNA. The most commonly used assay of the state of methylation of DNA utilizes methylation-sensitive restriction enzymes such as the isoschisomeric restriction enzymes pair: *Hpa*II and *Msp*I *(5)*. *Hpa*II cleaves the sequence CCGG when it is unmethylated at the internal cytosine residing in the dinucleotide CG sequence, whereas *Msp*I cleaves this site irrespective of its state of methylation. However, since the amount of demethylase activity assayed during a routine purification protocol is limited, nanogram quantities of methylated DNA are used as substrate, thus necessitating the use of Southern blotting, hybridization, and autoradiography. This might reduce the suitability of the assay for assaying multiple fractions following protein fractionation by chromatography. An additional limitation of the assay is that it is indirect and measures cleavage of DNA rather than methylation. The results might be masked by activity of nucleases and by other repair processes. The assay does not distinguish between true demethylation of DNA and repair of methylated cytosines and their replacement by nonmethylated cytosines.

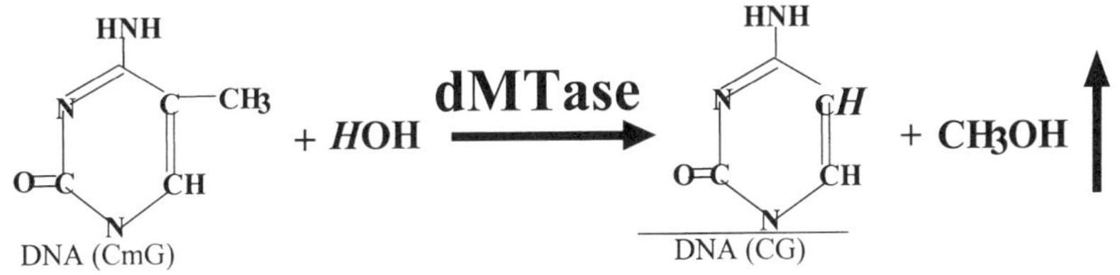

Fig. 2. The direct demethylation reaction. As shown in **ref. 4**, the methyl residue in methylated cytosine can react with water in a reaction catalyzed by the demethylase enzyme to form methanol and nonmethylated cytosines.

Another indirect assay of demethylation is bisulfite mapping *(6)*. This assay takes advantage of the fact that bisulfite treatment of DNA results in conversion of unmethylated cytosines into thymidines, whereas methylated cytosines are protected from this modification. Following treatment, specific sequences in the genome are amplified by polymerase chain reaction (PCR) using specific primers, the PCR fragments are subcloned and subjected to a sequencing analysis. Only methylated cytosines are detected in the cytosine-reaction lane. This technique in distinction from the restriction-enzyme analysis allows for visualization of methylated cytosines at a single-base resolution. However, while we have utilized in the past this technique to study the sequence specificity of demethylase and its processivity *(4,7)*, it is a labor-intensive procedure that is not suitable for identifying an active fraction following chromatographic fractionation, as discussed earlier. This assay, similar to restriction-enzyme analysis, measures methylation indirectly and does not distinguish between true demethylation and repair.

An assay that directly measures the conversion of methyl-CpG to CpG in a DNA substrate has been previously used by us to study demethylase and demonstrate its presence in mammalian cells *(4,8)*. This assay is discussed in a separate chapter in this volume. However, the CpG conversion assay is somewhat cumbersome and therefore unsuitable for routine purification procedure.

1.3. Measuring the Leaving Methyl Group by a Volatilization Assay

What distinguishes true demethylation from repair-based replacement of methylated cytosines is that the product of the reaction is a volatile methyl residue leaving as methanol *(4)*. The fact that the leaving group is volatile poses some obvious difficulties. However, simple scintillation cocktail-based assays were developed in the past to trap and measure leaving volatile residues such as methanol *(9)*. DNA is methylated with ^{3}H-methyl moieties. The release of the radioactive methyl residue is then measured in a gas-tight sealed environment, such as a scintillation vial containing a small compartment (a microfuge tube floating in the scintillation fluid) in which ^{3}H-methyl-DNA is incubated with the protein fraction and a larger compartment (the scintillation liquid). The released volatile ^{3}H-methanol diffuses from the smaller compartment into the joint space and will eventually be dissolved in the larger volume of scintillation liquid because of the volume ratio of the two liquid compartments. Methanol is thus transferred from the smaller volume of the reaction mixture to the larger volume of the scintillation liquid (**Fig. 3**). The counts detected by a scintillation counter correlate to the amount of methanol released in the reaction. This assay provides a clear and very sensitive method to test directly a large number of different chromatographic fractions for true demethylation activity.

1.4. Tissue Sources for Demethylase

One of the main obstacles hampering the attempts to identify the demethylase was identifying a tissue source that expresses high levels of demethylase activity. It was believed that demethylase should be expressed in early embryonal cells where active demethylation was known to take place *(1)*. Since the availability of sufficient amounts of early embryonal tissue is obviously limited, it is essential to use an alternative source. We have shown that cancer cells express relatively high levels of demethylase and we routinely use the human nonsmall-cell lung carcinoma (SCLC) cell line A549 (ATCC: CCL 185) *(4)*. An excellent source of material are A549-derived tumors, which are passaged as xenografts in nude mice *(4)*. In our experience these tumors bear high levels of demethylase activity.

1.5. Purification of Recombinant Demethylase Activity

We have recently cloned a cDNA from human cells that encodes demethylase activity *(10)*. The same cDNA was cloned independently by Bird's group and shown to have methylated-DNA binding activity *(11)* and to be part of a complex bearing histone deacetylase activity and was named Mdb2b *(12)*. Bird's group has failed to show that this cDNA bears demethylase activity *(12)*. We were not able to obtain active recombinant protein from bacterial-expression systems; we do not understand the reason for that. Perhaps proper folding is required or a specific post-translational modification that does not occur in bacterial cells is critical for demethylase activity. However we can purify a protein bearing demethylase activity from human embryonal kidney cells HEK 293 (ATCC: CRL 1573) that are transiently transfected with a plasmid expressing the cloned histidine-tagged demethylase cDNA under the control of a CMV promoter *(10)*. The 6xhistidine tag enables a powerful one step affinity purification of the recombinant demethylase by immobilized-metal affinity chromatography (IMAC) *(10)* using commercially available nickel-charged resins such as ProBond (Invitrogen).

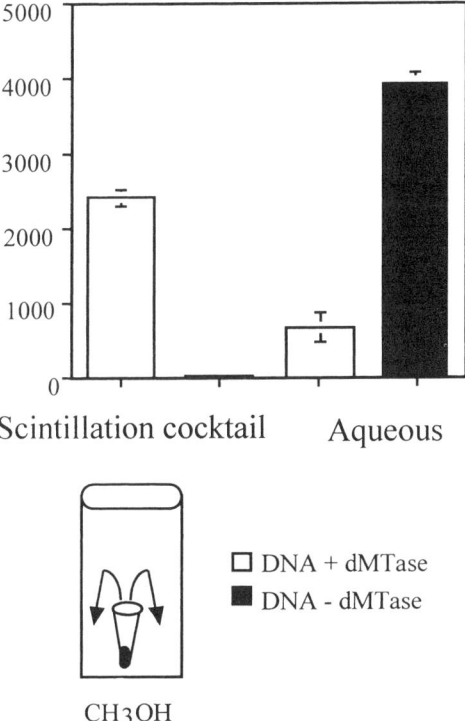

Fig. 3. The volatile demethylase assay. A 0.5-mL eppendorf tube containing ^{3}H-methyl-DNA and either demethylase or a control buffer are placed in scintillation vials containing scintillation cocktail. Upon demethylation, the released volatile ^{3}H-methanol diffuses from the smaller compartment into the joint space and will eventually be dissolved in the larger volume of scintillation liquid because of the volume ratio of the two liquid compartments. Methanol is thus transferred from the smaller volume of the reaction mixture to the larger volume of the scintillation liquid. The counts detected by a scintillation counter correlate to the amount of methanol released in the reaction. The aqueous phase containing the reaction mixture is later counted. The counts obtained for both fractions are graphed. In the presence of demethylase most counts are transferred to the liquid-scintillation cocktail, whereas in the absence of demethylase all the counts remain in the aqueous phase.

1.6. Specific Cautionary Points Regarding IMAC Purification of Recombinant Demethylase

Although IMAC purification of histidine-tagged protein is a straightforward and routine procedure, we have found out that some specific caveats associated with the idiosyncrasies of this protein can turn it into a frustrating experience. Histidine-tagged proteins are eluted from the IMAC column by a stepwise gradient with increasing concentrations of imidazole *(10)*. However, we found that imidazole inhibits demethylase activity at the low micromolar range. Extensive dialysis 3–4 times against 1000-fold volume of buffer is required to remove traces of imidazole. If this is not accomplished, the protein preparation is inactive. To our surprise, we also found that many routine dialysis procedures are ineffective since they result in either the adsorption of this highly charged protein to the dialysis membrane or in the formation of protein aggregates that are inactive. Traces of heavy metals are found in most regenerated cholesterol (RC) dialysis membranes. Since histidine-tagged proteins bind tightly to heavy metals, this might result in irreversible binding of the protein to the membrane. While several protocols are routinely used in different laboratories to decontaminate dialysis membrane from trace heavy metals, this procedure is not consistently effective in our hands. We strongly recommend the use of Biotech-grade membranes that are prepared from synthetic material and do not require heavy metals in the extraction procedure. Such membranes are commercially available from Spectrapor.

An additional caveat that has to be considered when dialyzing histidine-tagged proteins is the possibility that trace amounts of nickel are leached out of the Probond ImaC matrix. Since the rate of dialysis is related to the size of the molecule, the heavier imidazole will diffuse out of the membrane at a faster rate than nickel. This might result in a condition where nickel is still present in the dialysate in the absence of imidazole. The nickel atoms might precipitate complexes of nickel and multiple histidine-tagged protein monomers.

An additional point is that the protein and especially the diluted recombinant demethylase is extremely sensitive to deep-freezing and thawing. The purified diluted recombinant protein is inactivated by one round of freezing at –70°C and thawing. High-salt buffers can alleviate this problem (2–5 *M*). However the high salt might be an obstacle for further concentration of the protein and the enzymatic activity.

The highly charged demethylase protein adsorbs to the membranes found in some centrifugation-based concentration devices. We found that the recovery of demethylase from the Millipore concentrator is close to nil. The adsorption of demethylase to commercially available Millipore membrane is not prevented by the passivation protocol with polytryleneglycol (PEG), which is recommended by the manufacturer. We did not observe these problems with Microcon concentrators and we have observed enhancement of demethylase activity following Microcon concentration. We recommend using freeze-drying for concentration of the protein. The protein remains active using this procedure.

An additional impediment that might be encountered in the purification protocol is the presence of an unidentified inhibitor of demethylase in cellular extracts. Whereas correctly packed DEA-Sephadex A-50 (Pharmacia Biotech

0180-01) and Probond columns allow for separation of the demethylase activity from the contaminating inhibitory activity, the inhibitory activity tends to co-elute with the demethylase activity. The inhibitor tends to be more labile than the demethylase activity and therefore the demethylase activity is restored in the preparation after lengthy incubation at 37°C. Once, the activity is restored, it remains linear for a couple of days at 37°C. We have not yet characterized the nature of this inhibitor.

2. Materials
2.1. Buffers

1. Buffer L: 10 mM Tris-HCl, pH 7.5, 10 mM MgCl$_2$.
2. Buffer L containing 0.2 M NaCl.
3. Buffer L containing 5 M NaCl.
4. Buffer A: 10 mM Tris-HCl, pH 8.0, 1.5 M MgCl$_2$, 5 mM KCl, 0.5% Nonident 40.
5. Buffer B: 20 mM Tris-HCl, pH 8.0, 25% glycerol, 1.5 M MgCl$_2$, 0.4 M NaCl.
6. Binding buffer for IMAC: 20 mM sodium phosphate buffer, pH 7.8, 500 mM NaCl.
7. Wash buffer for IMAC: 20 mM sodium phosphate buffer, pH 6.0, 500 mM NaCl.
8. 3M Imidazole in Wash buffer, pH 6.8.
9. 10% TCA (trichloroacetic acid) solution.
10. 10 mg/mL Yeast t-RNA (Roche, #009495).
11. 2 mg/mL Sonicated herring sperm DNA (Roche #223646).
12. 10% Na-azide.

2.2. Materials

1. [^{3}H-CH^3]-S-Adenosyl-methionine (AdoMet) (5–15 mCi/mol) (New England Nuclear #155H).
2. SssI CG methylase (1 U/µL), the reaction buffer NEB2, and AdoMet (32 mM) New England Biolabs #2265).
3. Nap5 desalting columns (Pharmacia #17-0853-01).
4. DEAE Sephadex A50 weak anion-exchanger matrix (Pharmacia #17-0180-01).
5. Liquid-scintillation cocktail ScintiSafe 30% (Fisher #SX23-5).
6. GF/C filters (Fisher).
7. A vacuum-filtration device.
8. Phenol.
9. Chloroform.
10. Glass wool.
11. 3-mL Syringes.
12. Peristaltic pump.
13. Two-way stopcock (Biorad #732-8102).
14. Tubing for low-pressure chromatography (Biorad).
15. Tissue-culture medium DMEM (Gibco-BRL).

16. Fetal calf serum (FCS) (Gibco-BRL).
17. Phosphate-buffered saline (PBS).
18. Cell scraper.
19. 100-mm tissue-culture dishes (Falcon).
20. 5 ml Tissue homogenizer.
21. Gradient mixer.
22. Stirring bars.
23. Magnetic plates.
24. Scintillation counter.
25. Vaccum lyophilizer.
26. Dialysis membranes (SpectraPor CE #722-05716-000).
27. Closures for CE membranes (SpectraPor #722-05716-000).
28. Microfuge.
29. Low-speed centrifuge.
30. Mortar and pestle.
31. ProBond Resin (Invitrogen 46-0019).

3. Methods

3.1. Preparation of [^{3}H]-CH$_3$-plasmid DNA Substrate for Volatile Assays

1. Incubate 5 µg of plasmid DNA (such as bluescript SK from Stratagene) in a 100 µL total reaction volume containing 5µCi of [^{3}H]-CH$_3$-AdoMet (5 mCi/mmol), 5 µL SssI methylase (10 u/µL), and 20 µL of the buffer recommended by the manufacturer (NEB2) for 3 h at 37°C.
2. Verify incorporation of the radioactive methyl moiety by TCA precipitation. Remove a 1-µL aliquot from the reaction mixture and transfer into a 2.5-mL tube containing 2 mL 10% TCA and 20 µL (2 mg/mL) sonicated herring sperm DNA (Boehringer). Leave the sample on ice for 15 min. Pass the TCA precipitated sample through a GF/C filter using a vacuum-filtration apparatus. Wash the filter twice with 2 mL of cold 10% TCA on the vacuum-filtration device. Dry the filter under a heating lamp and place it in a 5-mL scintillation vial containing 2.5 mL scintillation cocktail. Count the samples in a liquid-scintillation counter. Using these labeling conditions, expect a reading of $1-2 \times 10^4$ dpm/µL.
3. Since the concentration of labeled AdoMet is below the K_m of the SssI methylase enzyme, one needs to chase the hot methylation reaction with cold AdoMet to fully methylate the substrate. Once the incorporation of labeled AdoMet onto the plasmid is verified, add 1 µL 3.2 mM nonradioactive AdoMet (NEB) and 5 µL SssI methylase to the reaction mixture and incubate it for an additional 3-h period.
4. To eliminate the possibility of contamination of the demethylation reaction with methylase used for preparation of the substrate, the methylated DNA is purified by phenol-chloroform extraction. Add 100 µL of phenol and 100 µL of chloroform to the reaction mixture. Following mixing of the phases using a vortex, the phases are separated by centrifugation at maximum speed in a

microfuge for 5 min. Transfer the aqueous phase to a new microfuge tube and avoid transfer of any of the organic phase.
5. The DNA is further purified by gel filtration twice to eliminate even minute amounts of residual AdoMet. Eliminating residual AdoMet is critical for the volatile demethylation assay. AdoMet is spontaneously degraded at room temperature to a volatile moiety. This will result in background counts that could mask the true demethylation products. Add 300 µL of buffer L to the reaction mixture. Wash a Nap5 column (Pharmacia) with 10 mL of buffer L. Load the 500 µL reaction mixture on the column. Elute the methylated DNA by applying a sequence of six 0.5-mL aliquots of buffer L and collect each 0.5-mL fraction in a separate microfuge tube. To determine which fraction contains the methylated DNA, transfer a 1-µL sample of each of the fractions into a tube containing 2 mL 10% TCA and measure the number of counts incorporated into DNA in each fraction as described in **Subheading 3.1., step 2**. Under these conditions, the [^{3}H-CH$_3$]-DNA elutes at the second and third fractions.
6. The fractions containing the DNA are concentrated. Add 10 µL tRNA, 50 µL 5 *M* NaCl, and 1 mL ethanol to each of the DNA-containing fractions. Incubate the samples in a dry-ice/ethanol bath for 15 min. Spin the samples in a microfuge at maximum speed for 15 min. Discard the supernatant, taking care not to disrupt the pellet, and dry the microfuge tube from residual ethanol with care using a Kimwipe tissue paper. Resuspend each of the two pellets in 0.25 mL buffer L and combine the two pellets. Load the combined 0.5 mL DNA solution on a fresh Nap5 column prepared as described in **Subheading 3.1., step 5** and elute the DNA containing fractions as described in **Subheading 3.1., step 4**. the DNA is typically eluted in two 0.5-mL fractions. Combine the fractions and store the the [^{3}H-CH$_3$]-labeled DNA at –20°C for future demethylase assays. The concentration of the DNA is 5 ng/µL and its specific activity is typically $1-2 \times 10^6$/µg.

3.2. Preparation of the DEAE-Sephadex Slurry and Packing of the Column

1. Mix 4 g of DEAE Sephadex A 50 matrix with 50 mL buffer L and allow it to swell by incubating the slurry at a 65°C water bath for 4 h. Add an additional aliquot of 50 mL of buffer L to the slurry and leave the slurry at 65°C for an additional 2 h. Repeat this procedure twice. For long-term storage of the slurry, add sodium azide to a final concentration of 0.5%. Store the slurry at 4°C.
2. We use a 5-mL syringe as a disposable column for purification of demethylase activity. Seal the bottom opening of the syringe with glass wool. Press the glass wool using the syringe plunger and verify that the glass wool forms a thin uniform layer. Attach a two-way stopcock to the to control liquid flow.
3. Wash the column with 10 mL buffer L (0.2 *M* NaCl) retaining 2 mL buffer L (0.2 *M* NaCl) in the syringe.
4. Add 2 mL of the DEAE slurry by slow pipetting into the buffer L in the column, and allow the matrix to settle.

5. Pierce a hole in the syringe plunger using a hot iron rod/needle and fix a 0.2 mL micropipet tip into it. Place the plunger in the syringe and fix chromatography tubing to the top end of the tip and connect it with a peristaltic pump.
6. Connect the tubing at the other end of the pump with a reservoir of buffer L (0.2 M NaCl). Turn on the pump and allow the buffer to drip into the syringe, filling it up to the top and retaining the stopcock in the closed position.
7. Open the stop cock and press the plunger to about 0.5 cm from the top of the syringe.
8. Pass 20 mL of buffer L (0.2 M NaCl) through the column at a flow rate of 1 mL/min to equilibrate the column. Stop the peristaltic pump and close the stopcock. The column is ready for loading.

3.3. Preparation of Nuclear Extracts from Tumor Samples and Cell Lines

1. Forty plates of A549 cells are grown to 90% confluency in Dulbecco's Modified Eagle's Medium (DMEM) medium containing 10% FCS. To harvest the cells, remove the medium and apply 5 mL cold PBS. Scrape the cells off the tissue-culture dish using a rubber cell scraper and transfer the cells to four 50-mL tubes. Spin the cells at 1.5×10^3 rpm for 10 min in a low-speed centrifuge. Remove the PBS by aspiration.
2. To prepare demethylase from A549 tumors that are passaged as xenografts in nude mice, crush 1 g of a tumor in a mortar and pestle in the presence of liquid nitrogen. Resuspend the tumor paste in buffer A and homogenize it using a 5-mL glass homogenizer.
3. Resuspend each cell pellet in 0.5 mL of buffer A and leave the suspension on ice for 15 min. Do not use any protease inhibitors since most of the routinely used protease inhibitors inhibit demethylase-activity preparation.
4. Spin the suspension at 2×10^3 rpm in a microfuge for 15 min at 4°C to isolate the nuclei. Discard the supernatant and resuspend each nuclear pellet in 0.5 mL of buffer A to remove residual cytosolic proteins. Keep the nuclear suspension on ice for 10 min and spin the samples in a microfuge at 2×10^3 rpm for 15 min and discard the supernatant.
5. The nuclear proteins are eluted in buffer B, which contains 0.4 M NaCl. Resuspend each nuclear pellet in 0.2 mL of buffer B and incubate the suspension at 4°C for 20 min. Mix the sample carefully by stirring with a micropipet tip.
6. Spin the cells at 15,000 rpm for 30 min in a microfuge. Transfer the supernatant containing the nuclear extract to a fresh tube. Spin again at the same speed for additional 30 min. The nuclear extract is ready for loading on the DEAE column. Determine the concentration of protein in the extract using the BioRad protein determination kit as described by the manufacturer. We typically obtain 7 mg of nuclear proteins from 40 plates of A549 cells and 14 mg of nuclear proteins from 1 g of A549 tumors.
7. Dilute the nuclear extract to a final concentration of 0.2 M NaCl by adding 1 mL of buffer L to 0.8 mL of nuclear extract. Open the stopcock, load the diluted

nuclear extract on the column, and collect the flow-through. Wash the column by passing 10 mL of buffer L (0.2 M NaCl) through the column at a rate of 0.2 mL/min and collect 1.5-mL fractions.
8. Demethylase is eluted from the column by a NaCl gradient. Use a 10-mL gradient mixer. Verify that the opening between the two chambers is closed and add 5 mL of buffer L (5 M NaCl) in the left cell of the gradient mixer and 5 mL of buffer L (0.2 M NaCl) in the right cell. Place a stirring magnetic bar in the right cell, place the gradient mixer on a stirring plate and attach it with the tubing, connecting it with the column through the peristaltic pump. Open the opening between the chambers and run the gradient through the column at a flow rate of 1 mL/min and collect twenty 0.5-mL fractions. Perform the chromatography steps in a cold room and keep the samples at all time at 4°C. Remove 20-µL aliquots of all the collected fractions for demethylase assay and freeze the samples at –20° C. The samples remain active for at least 4 wk at –20°C.

3.4. Determination of Demethylase Activity

1. Perform the demethylation reaction in 0.5-mL microfuge tubes.
2. Add 20 µL buffer L, 2.5 µL [^{3}H-CH$_3$]-SK DNA (5 ng/µL, 1000 dpm/µL), and 20 µL of either buffer L for the control sample or 20 µL of each of the different test samples into each tube.
3. Add 2 mL liquid-scintillation cocktail into 5-mL polypropylene scintillation vials. Place the 0.5-mL microfuge tubes containing the demethylation reaction mixture in the scintillation vials and leave their caps open. Place the tubes with special care to avoid spilling of reaction mixture into the scintillation cocktail, which could result in false-positives. Place the caps on the scintillation vials and seal them tightly. Allow the reaction products to volatilize for 24 h at 37°C.
4. The radioactive methyl group leaves as methanol and diffuses into the scintillation cocktail. Count the scintilation vials in a scintillation counter. Control samples do not register any counts above the background of the counter (0–20 dpm). In samples that bear demethylase activity, radioactive methanol diffuses into the scintillation liquid, resulting in registered counts. Under these conditions 20–50% of the DNA is demethylated by the active fractions, resulting in volatilization of 500–1000 dpm (*see* **Fig. 4**).
5. The active fractions are pooled and concentrated by lyophilization using a standard lyophilizer. Do not use membrane concentrators since the protein tends sticks to the membranes. The concentrated preparation could be used for further biochemical studies.

3.5. Purification of Recombinant Demethylase from Transiently Transfected HEK Cells

1. We have previously published an expression vector encoding a histidine-tagged human demethylase cDNA. This construct, which is driven by a cytomegalovirus (CMV) promoter, could be expressed following transient transfection into mammalian cells such as the human embryonal kidney cell line HEK 293 (ATCC:

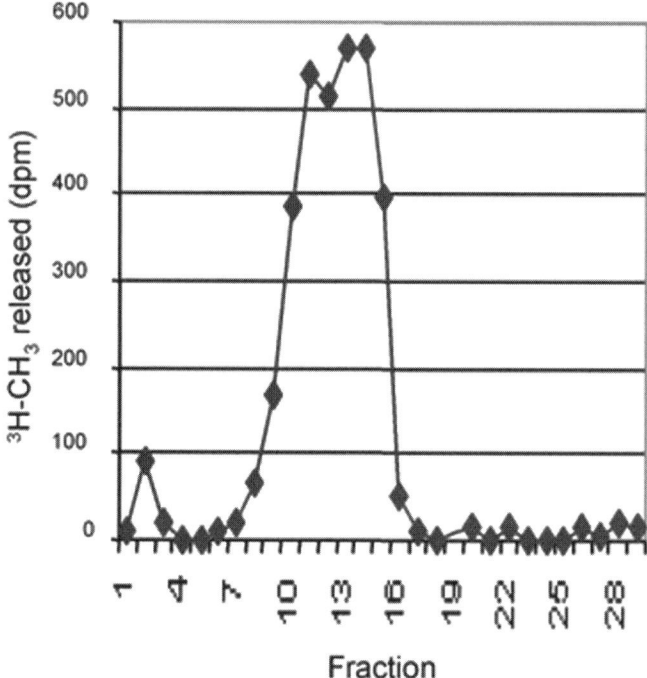

Fig. 4. Chromtographic fractionation of demethylase activity from A549 cells. Seven mg of A549 nuclear extract were loaded on a DEAE sephadex A50 column as described. The different fractions were collected as described. Aliquots of the diferent fractions were subjected to a demethylase volatilization assay as described in **Subheading 3.** and were counted after 32 h.

CCL: 185). The plasmid could be introduced by any of the established transfection procedures, which will not be discussed here. We use the well-established calcium phosphate precipitation method *(13)*. For a single demethylase preparation, we transfect 10 plates, each with 10 µg of the plasmid his-demethylase as previously described *(10)*.

2. Forty-eight h after transfection, aspirate the medium and add 5 ml of ice-cold PBS per plate. Disrupt the attachment of the cells to the plate surface using a 10-mL pipettor and transfer the cells into a 50-mL centrifuge tube. Spin the cells at 1,500 rpm at 4°C for 10 min. Aspirate the PBS and resuspend the cells in 1 mL of binding buffer and transfer into an eppendorf tube. Do not add protease inhibitors.

3. Prepare a dry-ice ethanol bath and place the eppendorf tube containing the HEK cell suspension in it for 5 min. Transfer the frozen sample into a 37°C bath and leave it to thaw for 5 min. Repeat the freeze-thaw cycle once. To reduce the viscosity of the extract shear the DNA contained in the extract by passing the cell extract four times through a 18.5-gauge needle using a 3-mL syringe. Pellet

the cell debris by centrifugation at 15,000 rpm for 30 min at 4°C. Transfer the supernatant into a fresh tube.
4. Histidine-tagged proteins could be purified by IMAC. The histidine tag has high affinity to heavy metals such as nickel. Different immobilized heavy-metal matrices are commercially available. We use the Probond matrix from Invitrogen. We use a 1-mL tip as our column. Cut the tip of the pipet tip and seal the opening by packing glass-wool as described in **Subheading 3.2., step 2**. Attach a two-way stopcock to the tip to control the flow. Wash the glass wool with 2 mL binding buffer and leave 0.3 mL of buffer in the column. Mix the Probond slurry and pipet slowly 0.5 mL of slurry into the buffer in the column. Allow the buffer to drip and wash the column with 3 mL of binding buffer at a flow rate of 0.2 mL/min. Close the stopcock when 0.2 mL of buffer are left in the column. This is important to avoid drying of the matrix. The column is now ready.
5. Load the extract on the column and collect the flow-through. Wash the column with 3 mL binding buffer and follow the wash with 3 mL washing buffer at a flow rate of 0.2 mL/min. Collect 1.5-mL fractions. Elution of the nickel-bound column is performed by a stepwise gradient of imidazole (50 mM–1 M). Apply sequentially 0.5 mL of 50 mM, 200 mM, 350 mM, 500 mM, 700 mM, and 1 M imidazole washing buffer solutions on the column and collect each fraction separately. Leave the fractions on ice.
6. Soak a sufficient length of dialysis tubing (Spectrapor CE membranes 15 mm) in 400 mL of a 1 mM EDTA solution for 30 min at room temperature. Rinse the membrane three times in double-distilled buffer. Keep the membranes soaked in water and avoid drying of the membrane.
7. Seal one end of the dialysis tubing with the special clamps provided by the manufacturer and transfer the different imidazole fractions to separate tubes. Seal the dialysis bags with a second clamp and mark each bag clearly by marking the clamp with a water-resistant marker. Place the three bags in a beaker containing 1 L of buffer L, place a magnet in the beaker, and place it on a stirring plate in the cold room. Leave the samples to dialyze for 6 h and repeat the dialysis step twice.
8. Following three dialysis steps, transfer the contents of the bags into eppendorf tubes and store at –20°C. To assay the samples follow the steps described in **Subheading 3.4.**

Acknowledgments

The research reported here was supported by grants from the Medical Research Council of Canada and the National Cancer Institute of Canada to MS.

References

1. Kafri, T., Gao, X., and Razin, A. (1993) Mechanistic aspects of genome-wide demethylation in the preimplantation mouse embryo. *Proc. Natl. Acad. Sci. USA* **90,** 10,558–10,562.

2. Razin, A., Szyf, M., Kafri, T., Roll, M., Giloh, H., Scrapa, S., et al. (1986) Replacement of 5-methylcytosine by cytosine: a possible mechanism for transient DNA demethylation during differentiation. *Proc. Natl. Acad. Sci. USA* **83,** 2827–2831.
3. Weiss, A., Keshet, I., Razin, A., and Cedar, H. (1996) DNA demethylation in vitro: involvement of RNA. *Cell* **87,** 709–718.
4. Ramchandani, S., Bhattacharya, S. K., Cervoni, N., and Szyf, M. (1999) DNA methylation is a reversible biological signal. *Proc. Natl. Acad. Sci. USA* **96,** 6107–6112.
5. Waalwijk, C. and Flavell, R. A. (1978) DNA methylation at a CCGG sequence in the large intron of the rabbit b-globin gene: tissue specific variations. *Nucleic Acids. Res.* **5,** 4631–4641.
6. Clark, S. J., Harrison, C. L., Paul, C. L., and Frommer, M. (1994) High sensitivity mapping of methylated cytosines. *Nucleic Acids Res.* **22,** 2990–2997.
7. Cervoni, N., Bhattacharya, S. K., and Szyf, M. (1999) DNA demethylase is a processive enzyme. *J. Biol. Chem.* **274,** 8363–8366.
8. Szyf, M., Theberge, J., and Bozovic, V. (1995) Ras induces a general DNA demethylation activity in mouse embryonal P19 cells. *J. Biol. Chem.* **270,** 12,690–12,696.
9. Terwilliger, T. C., Bogoonez, E., Wang, E. A., and Koshland Jr., D. E. (1983) Sites of methyl esterification on the aspartate receptor involved in bacterial chemotaxis. *J. Biol. Chem.* **258,** 9608–9611.
10. Bhattacharya, S. K., Ramchandani, S., Cervoni, N., and Szyf, M. (1999) A mammalian protein with specific demethylase activity for mCpG DNA. *Nature* **397,** 579–583.
11. Hendrich B. and Bird A. Identification and characterization of a family of mammalian methyl-CpG binding proteins. (1998) *Mol. Cell. Biol.* **18,** 6538–6547.
12. Ng, H. H., Zhang, Y., Hendrich, B., Johnson, C. A., Turner, B. M., Erdjument-Bromage, H., et al. (1999) MBD2 is a transcriptional repressor belonging to the MeCP1 histone deacetylase complex. *Nature Genet.* **23,** 58–61.
13. Ausaubel, F. M., Brent, R., Kingston, R. E., Moore, D. D., Smith, J. A., Seidman, J. G., and Stairwell, K. (eds.) (1988) *Current Protocols in Molecular Biology.* John Wiley & Sons, New York.

Index

A

Abnormally methylated DNA fragment characterization
  MSRF
    methods, 48
Adaptor ligation
  MCA amplicons, 104
  RLGS, 68
Altered methylation status confirmation
  methods
    MSRF, 48–49
Amplicon generation
  DMH, 93–97
    Cot-1 subtraction, 95–96
    linker ligation, 94
    methylation-sensitive restriction and amplification, 97
    MseI digestion, 93
Amplicons
  DMH, 88–89
  DNA
    testing steps, 99
  MCA, 104–105
  ovary, 89
  tumor, 89
Array hybridization
  DMH, 97–98

B

Binding buffer
  Cot-1 subtraction, 96

BioRex 70
  Xenopus laevis extract fractionation, 137
Biotech-grade membranes, 168
Bisulfite-induced deamination
  cytosines, 29
Bisulfite mapping, 165
Bisulfite sequencing
  LM-PCR, 31
Bisulfite-treated DNA
  PCR amplification, 77
Bisulfite treatment
  COBRA
    materials, 82
    methods, 83–84
Bisulphite-assisted genomic sequencing, 143–153
  improvements, 144
  materials, 145
    embedding material, 145
    PCR purification and cloning, 145
  methods, 145–152
    bisulphite reaction chemistry, 145–146
    cell preparation, 147–149
    cloning and sequencing, 150–152
    drawbacks, 152
    isolated DNA, 149
    PCR optimization, 150
    primer design, 150
    principles, 146–147
    problems, 151

177

BssHII enzyme, 53
BstUI enzyme, 99
Buffers
   DNA cytosine-5 methylation
      RP-HPLC, 18
   DNA demethylase, 158
   extraction, 169

C

Cancer
   CpG islands, 113
   DMH, 87
   DNA methylation, 4
CA-RLGS, 54
CDK6
   gene, 54
CGI clones
   testing steps, 99
CGI genomic library, 87–88
Chart recorder
   cytosine-5 methylation, 23–26
Chemical DNA sequencing, 29–39
   C reaction, 36
   G+A reaction, 35
   gel analysis sequencing
      reaction products, 37–38
   G-reaction, 35
   LM-PCR, 36–37
   materials, 31–34
   methods, 34–38
   T+C reaction, 36
Chromatography
   MeCp2-containing deacetylase, 135
Chromosome-assigned RLGS (CA-RLGS), 54
Cloned DNA digestion
   MBD columns, 120

Cloning
   bisulphite-assisted genomic sequencing, 150–152
   MSRF
      materials, 45
COBRA. *see* Combined bisulfite restriction analysis (COBRA)
Colony PCR
   CpG island array preparation
      DMH, 90–91
      master mix, 91
Combined bisulfite restriction analysis (COBRA), 71–85
   electroblotting, 81
   experimental design, 74–82
      PCR primer design, 75–77
      probe design, 78
      restriction-enzyme choice, 77–78
      sequence manipulation, 74–75
   hybridization, 81
   materials, 81–83
      bisulfite treatment, 82
      DNA extraction, 82
      DNA isolation, 81–82
      electroblotting, 83
      hybridization, 83
      PCR, 82
      polyacrylamide gel electrophoresis, 82–83
      restriction enzyme digestion, 82
   methods, 83–85
      bisulfite treatment, 83–84
      DNA clean up, 84
      DNA isolation, 83
      electroblotting, 84–85
      hybridization, 84–85
      PCR amplification, 84
      polyacrylamide gel electrophoresis, 84

Index   179

prehybridization, 84–85
restriction enzyme digestion, 84
washing, 84–85
PCR, 80
polyacrylamide gel
    electrophoresis, 80–81
protocol outline, 73
restriction enzyme digestion, 80
restriction enzymes, 78
sodium bisulfite techniques,
    72–73
sodium bisulfite treatment, 78–80
Computer-stimulated bisulfite
    conversion, 74
Contaminating RNA
    incomplete hydrolysis, 26
Cot-1
    biotin labeling, 95
Cot-1 hybridization
    CpG island array preparation
        DMH, 92–93
Cot-1 subtraction
    amplicon generation
        DMH, 95–96
    binding buffer, 96
    documentation, 98
CpG
    percent methylation
        estimation, 11–12
CpG dinucleotides, 1
    highlight, 76
CpG island array preparation
    DMH, 90–93
        colony PCR, 90–91
        Cot-1 hybridization, 92–93
        methylation-sensitive
            restriction digest, 91–92
CpG island clones
    potential analysis, 126–127

CpG island libraries
    genomic DNA, 121
    large genomic clones, 121
    MBD columns, 112–113
CpG island promoters
    hypermethylation, 4
CpG islands, 2, 3, 43
    bias, 55
    estimated number, 127
    MBD column
        materials, 113–114
        methods, 114–121
    methylation, 4
    methylation-free, 111–112
    methyl-CpG binding column,
        111–127
    primers, 108–109
    restriction enzymes, 112
    RLGS, 53
Cyclin-dependent kinase-6 (CDK6)
    gene, 54
Cytosine
    sodium bisulfite conversion, 78–79
Cytosine-5 methylation
    enzymes, 2
    RP-HPLC, 17–26
        computer analysis, 23
        materials, 18
        method, 18–26
        reproducibility tests, 26
Cytosine residue methylation
    genomic sequencing protocol, 144
Cytosine residues
    methylation, 43
Cytosines
    bisulfite-induced deamination, 29

D

DCMP, 17
  computerized data acquisition, 24
Deacetylase assay
  MeCp2-containing deacetylase, 137
DEAE-Sephadex slurry
  DNA demethylase extraction, 171–172
Demethylase
  definition, 163
Demethylation assay, 160
Demethylation reaction
  product detection, 163–165
2'-deoxycytidine-5'-monophosphates (dCMP), 17
2'-deoxyribonucleotide-5'-monophosphates (dNMPs)
  separation, 20–22
Deoxyribonucleotide standards, 25
Developing tanks
  TLC, 13
Differentially methylated plasmid DNAs
  MBD column calibration, 118–119
Differential methylation hybridization (DMH), 131
  amplicons, 88–89
  array hybridization, 97–98
  CpG island arrays, 87–99
  genomic DNA sequences, 87
  materials, 89–90
  methods, 90–98
    amplicon generation, 93–97
    CpG island array preparation, 90–93
  representative results, 89
  restriction enzymes, 88
  schematic flowchart, 88

DMH. *see* Differential methylation hybridization (DMH)
DNA
  hydrolysis, 19
  measurement
    first-dimension electrophoresis, 60–61
  postreplicative methylation
    cytosine C5 position, 143
  preparation, 18–29
  satellite II, 4
  satellite III, 4
DNA amplicons
  testing steps, 99
DNA-analysis program
  restriction maps, 77
DNA-binding proteins
  Southwestern analysis, 132
DNA clean up
  COBRA
    methods, 84
DNA demethylase
  in vitro measurement, 155–160
    alternative assays, 155–156
    demethylation assay, 160
    materials, 156–158
    methods, 158–160
    solutions and buffers, 158
DNA demethylase extraction
  [^{3}H]-CH$_3$-plasmid DNA substrate, 170–171
  mammalian cells, 163–175
  materials, 169–170
  methods, 170–175
    activity determination, 173
    DEAE-Sephadex slurry, 171–172
    [^{3}H]-CH$_3$-plasmid DNA substrate, 170–171
    nuclear extracts, 172–173

PEG, 168
scintillation counter, 166
tissue sources, 166
volatilization assay, 165–166
DNA digests
    MSRF
        methods, 46–47
DNA elution
    MSRF
        materials, 45
        methods, 47–48
    RLGS
        methods, 66–68
DNA extraction
    COBRA
        materials, 82
    cultured cells, 50
    MSRF
        materials, 44
DNA fragment cloning
    RLGS
        materials, 56–57
        methods, 65–68
DNA isolation
    chemical DNA sequencing, 34–35
    COBRA
        materials, 81–82
        methods, 83
    MSRF
        methods, 46
DNA methylation
    analysis
        single nucleotide resolution, 29
    indirect detection, 71
    protein binding, 2
    role, 3–5
DNA methyltransferase 3a
        (Dnmt3a), 2
    deletion, 3

DNA methyltransferase 3b
        (Dnmt3b), 2
    deficiency, 4
    deletion, 3
DNA methyltransferase 1 (Dnmt1), 2
    targeted deletion, 3
DNA preparation
    MBD columns, 120
DNA strand sequences
    COBRA, 74–75
DNMPs
    separation, 20–22
Dnmt, 2–4
Dot-blot analysis
    MCA
        filter preparations, 105
        hybridization, 105
Double-stranded oligomers
    synthesis
        ^{32}P-labeled [dC^{32}pdG]n,
            158–159
        ^{32}P-labeled [mdC^{32}pdG]n,
            158–159
Double-stranded plasmids
    ^{32}P-labeled [dC^{32}pdG]n
        synthesis, 159
    ^{32}P-labeled [mdC^{32}pdG]n
        synthesis, 159
Down's syndrome, 54
Dual channel chart recorder
    cytosine-5 methylation, 21

E

EagI enzyme, 53
Electroblotting
    COBRA, 81
        materials, 83
        methods, 84–85
End-labeled plasmid DNAs
    MBD column calibration, 119

Enzyme
    MspI, 164
Enzyme combinations
    RLGS, 60
Enzymes
    BssHII, 53
    BstUI, 99
    cytosine-5 methylation, 2
    digestion
        COBRA materials and
            methods, 78, 80, 82, 84
    DNA cytosine-5 methylation
        RP-HPLC, 18
    EagI, 53
    experimental design
        COBRA, 77–78
    HhaI, 155
    HpaII, 155, 164
    landmark, 53
    methylation-sensitive restriction
        DNA digestion, 143
Epigenetic gene silencing, 111

F

First-dimension electrophoresis
    RLGS
        DNA measurement, 60–61
        materials, 56
        methods, 60–62
First-dimension gel
    RLGS, 61–62

G

Gel analysis sequencing
    reaction products
        chemical DNA sequencing,
            37–38
Gel set-up
    RLGS
        materials, 56

Gene
    CDK6, 54
Genomic DNA
    CpG island libraries, 121
    digestion
        MBD columns, 120
        MCA amplicons, 104
    enzymatic processing RLGS
        materials, 56
        methods, 59–60
    sequences
        DMH, 87
Genomic sequencing protocol
    cytosine residue methylation, 144
Global hypomethylation, 4

H

[^{3}H]-CH$_3$-plasmid DNA substrate
    DNA demethylase extraction,
        170–171
Heavy metals
    regenerated cholesterol dialysis
        membranes, 168
Heparin
    Xenopus laevis extract
        fractionation, 139–140
HhaI enzyme, 155
High-pressure liquid chromatograph
        (HPLC)
    poor performance, 26
    solutions, 18
High-throughput DNA array
        technologies, 87
Histidine-tagged proteins
    deep-freezing and thawing, 168
    dialyzing, 168
Histone acetylation
    in vitro
        MeCp2-containing
            deacetylase, 136–137

Histone deacetylase assay
  MeCp2-containing deacetylase, 135
HMBD protein
  nickel-agarose resin coupling
    MBD column, 116–117
  preparation
    MBD column, 115
  purification
    MBD column, 116
HpaII enzyme, 155, 164
HPLC. *see* High-pressure liquid chromatograph (HPLC)
Human cerebellum DNA
  RLGS profile, 67
Hybridization
  COBRA, 81
    materials, 83
    methods, 84–85

I

ICF syndrome, 1, 4
Imprinted genes
  DNA methylation, 3
Inactive X chromosome genes
  DNA methylation, 3
In-gel digest
  RLGS
    materials, 57
    methods, 62
Interrogating probes
  DMH, 88–89
In vitro histone acetylation
  MeCp2-containing deacetylase, 134–135
In vivo footprinting, 31
Isocratic reverse-phase high-pressure liquid chromatography (RP-HPLC), 20–24
  mobile *vs.* solid phase, 20

Isolated DNA
  bisulphite-assisted genomic sequencing, 149

L

Landmark enzymes, 53
Large genomic clones
  CpG island libraries, 121
Ligation PCR
  test gel, 98
Linker ligation
  amplicon generation
    DMH, 94
Litigation-mediated PCR (LM-PCR), 29–32
  bisulfite sequencing, 31
Loss of function, 4

M

Mammalian cells
  DNA demethylase activity, 163–175
Mammalian DNA methylation, 1
Mammalian gene
  methylated cytosine detection, 32
Maxam-Gilbert procedure, 35
MBD. *see* Methyl-CpG binding domain (MBD)
MCA. *see* Methylated CpG island amplification (MCA)
5mdCMP
  computerized data acquisition, 24
  measurement, 17
MeCP. *see* Methyl-CpG binding activity (MeCP)
Methylated CpG island amplification (MCA), 101–103
  advantages and disadvantages, 103
  amplicons, 104–105
    adapter ligation, 104

    genomic DNA digestion, 104
    PCR amplification, 104–105
  materials, 103–104
    oligonucleotides, 104
    RDA and cloning PCR
      products, 103
  methods, 104–108
    dot-blot analysis, 105
    MCA amplicons, 104–105
  outline, 102
  positive and negative controls, 109
  RDA, 105–108
    competitive hybridization, 107
    differentially methylated
      sequences
      identification, 109
    driver amplicon adaptor
      removal, 106–107
    outline, 105–106
    second-round subtraction, 108
    selective amplification, 107–108
    tester amplicon adaptor
      change, 107
Methylated cytosine
  detection
    mammalian gene, 32
Methylated DNA binding
  MBD column calibration, 119–120
Methylation-sensitive restriction
    digest
  CpG island array preparation
    DMH, 91–92
Methylation-sensitive restriction
    enzymes
  DNA digestion, 143
Methylation-sensitive restriction
    fingerprinting (MSRF), 43–50
  banding pattern outcomes, 49

  materials, 44–46
    cloning, 45
    DNA elution, 45
    DNA extraction, 44
    PCR amplification, 44–45
    polyacrylamide gel
      electrophoresis, 45
    restriction enzyme digests, 44
    sequencing, 46
    Southern blotting, 45
  methods, 46–50
    abnormally methylated DNA
      fragment
      characterization, 48
    altered methylation status
      confirmation, 48–49
    DNA digests, 46–47
    DNA elution, 47–48
    DNA isolation, 46
    PCR amplification, 47
    polyacrylamide gel
      electrophoresis, 47
    results interpretation, 49–50
Methylation-specific PCR, 72
Methyl-CpG binding activity
    (MeCP), 2
Methyl-CpG binding column
  CpG islands, 111–127
Methyl-CpG binding domain
    (MBD), 2
Methyl-CpG binding domain 2
    (MBD2), 2–3
Methyl-CpG binding domain 4
    (MBD4), 3
Methyl-CpG binding domain
    (MBD) columns, 113–114
  applications, 121, 125–127
  calibration, 123–124
    materials, 114
    methods, 114–120

CpG island libraries, 112–113
DNA preparation, 124–125
preparation, 122
   materials, 113–114
   methods, 115–117
running, 122–123
   materials, 114
   methods, 117–118
Methyl-CpG binding protein (MeCP), 1
Methyl-CpG binding protein 1 (MeCP1), 131
Methyl-CpG binding protein 2 (MeCP2), 2, 131
  histone deacetylase complex, 132
  mutations, 5
Methyl-CpG binding protein 2 (MeCP2)-containing deacetylase
  histone deacetylase assay, 135
  materials, 132–135
    chromatography, 135
    histone deacetylase assay, 135
    oocyte extract preparation, 135
    Southwestern assay, 134
    Southwestern oligo preparation, 132–134
    in vitro histone acetylation, 134–135
  methods, 136–140
    deacetylase assay, 137
    oocyte extract fractionation, 137–140
    oocyte extract preparation, 137
    probe preparation, 136
    Southwestern blotting, 136
    in vitro histone acetylation, 136–137
  purification
    recombinant yeast Hat1 p, 140
  Xenopus laevis, 131–140

5-Methylcytosine
  hydrolytic deamination, 2
5-Methyl-2'-deoxycytidine-5'-monophosphate (5mdCMP)
  measurement, 17
Methyl-specific transcriptional repression, 131
Methyltransferase, 1
Molecular hybridization
  disadvantages, 71
MonoQ Sepharose
  Xenopus laevis extract fractionation, 138
MseI digestion
  amplicon generation
    DMH, 93
MspI enzyme, 164
MSRF. *see* Methylation-sensitive restriction fingerprinting (MSRF)
Ms-SNuPE, 144

N

Nearest-neighbor analysis, 9–15
  equipment, 11
  materials, 10–11
  methods, 11–15
  procedure, 9–10
  TLC, 12–15
Nickel-agarose resins, 122
Normal amplicon
  ovary, 89
NotI/EcoRV fragments
  NotI restriction trapper, 65–66
NotI restriction trapper
  NotI/EcoRV fragments, 65–66
Novel imprinted genes, 54
Nuclear extracts
  DNA demethylase extraction, 172–173

Nucleotide-3'-monophosphates
  TLC, 15
Nucleotides
  chart recorders, 25–26
  detection, 23
Nucleotide standards
  DNA cytosine-5 methylation
    RP-HPLC, 18

O

Oligomers
  synthesis
    ^{32}P-labeled [dC^{32}pdG]n
      double-stranded, 158–159
    ^{32}P-labeled [mdC^{32}pdG]n
      double-stranded, 158–159
Ovary
  normal amplicon, 89
  tumor amplicon, 89

P

PCNA, 2
PCR amplification, 143–144
  bisulfite-treated DNA, 77
  bisulphite-assisted genomic
    sequencing, 150
  COBRA, 80
    materials, 82
    methods, 84
  disadvantages, 71, 144
  MCA amplicons, 104–105
  MSRF
    materials, 44–45
    methods, 47
  RLGS, 68
PCR-based global methylation-
    analysis
  vs. RLGS-M, 55
PCR primer design
  COBRA, 75–77

PCR products
  MCA, 103
PCR purification and cloning
  bisulphite-assisted genomic
    sequencing, 145
PEG
  DNA demethylase, 168
P16/INK4 gene, 54
^{32}P-labeled [dC^{32}pdG]n double-
    stranded oligomers
  synthesis
    DNA demethylase, 158–159
^{32}P-labeled [dC^{32}pdG]n double-
    stranded plasmids
  synthesis
    DNA demethylase, 159
^{32}P-labeled [mdC^{32}pdG]n double-
    stranded oligomers
  synthesis
    DNA demethylase, 158–159
^{32}P-labeled [mdC^{32}pdG]n double-
    stranded plasmids
  synthesis
    DNA demethylase, 159
Plasmid
  synthesis
    ^{32}P-labeled [dC^{32}pdG]n
      double-stranded, 159
    ^{32}P-labeled [mdC^{32}pdG]n
      double-stranded, 159
Plasmid DNA
  [^{3}H]-CH$_3$
    DNA demethylase extraction,
      170–171
    MBD column calibration
      differentially methylated,
        118–119
      end-labeled, 119

Polyacrylamide gel electrophoresis
    COBRA, 80–81
        materials, 82–83
        methods, 84
    MSRF
        materials, 45
        methods, 47
Polytryleneglycol (PEG)
    DNA demethylase, 168
Prehybridization
    COBRA
        methods, 84–85
Primer ligation
    CMH
        reaction mix, 95
Primers
    bisulphite-assisted genomic
        sequencing, 150
    PCR amplification
        bisulfite-treated DNA, 77
Probe
    design
        COBRA, 78
    preparation
        MeCp2-containing
            deacetylase, 136
        single-stranded
            preparation, 38
Proliferation cell nuclear antigen
    (PCNA), 2
Purified fragments
    two-dimensional separation
        RLGS, 66

R

Radiolabeled NotI sites, 53
RDA
    MCA, 103, 105–108
        competitive hybridization, 107
        differentially methylated
            sequences
            identification, 109
        driver amplicon adaptor
            removal, 106–107
        outline, 105–106
        second-round subtraction, 108
        selective amplification, 107–108
        tester amplicon adaptor
            change, 107
Reagents
    DNA cytosine-5 methylation
        RP-HPLC, 18
    nearest-neighbor analysis, 10–11
Recombinant DNA demethylase
    IMAC purification, 167–169
    purification, 166
        transiently transfected HEK
            cells, 173–175
Recombinant yeast Hat1 p
    MeCP-2 containing deacetylase
        purification, 140
Regenerated cholesterol dialysis
    membranes
    heavy metals, 168
Restricting genomic DNA
    DMH
        mix, 94
Restricting PCR-amplification tags
    CpG island array preparation
        master mix, 92
Restriction enzymes
    COBRA, 78
    COBRA choice, 77–78
    COBRA digestion, 80
        materials, 82
        methods, 84
    CpG islands, 112
    digests, 71
        MSRF materials, 44

DMH, 88
DNA digestion, 143
    site identification, 77
Restriction landmark genome
    scanning (RLGS), 53–69
    adaptor ligation, 68
    materials, 55–58
        DNA fragment cloning, 56–57
        first-dimension
            electrophoresis, 56
        gel set-up, 56
        genomic DNA enzymatic
            processing, 56
        genomic DNA isolation, 55
        in-gel digest, 57
        RLG profile analysis, 56
        second-dimension
            electrophoresis, 57
    methods, 58–59
        DNA fragment cloning, 65–68
        first-dimension
            electrophoresis, 60–62
        gel set-up, 60–62
        genomic DNA enzymatic
            processing, 59–60
        genomic DNA isolation, 58–59
        in-gel digest, 62
        RLG profile analysis, 64–65
        second-dimension
            electrophoresis, 63–64
    PCR amplification, 68
    vs. PCR-based global
        methylation-analysis, 55
    profile
        human cerebellum DNA, 67
    profiles
        direct visual assessment, 64
Restriction maps
    DNA-analysis program, 77

Rett syndrome, 1, 5
Reverse-phase high-pressure liquid
    chromatography (RP-HPLC),
    20–24
    DNA cytosine-5 methylation,
        17–26
RLGS. see Restriction landmark
    genome scanning (RLGS)
RMCA primers, 109
RP-HPLC, 20–24
    DNA cytosine-5 methylation,
        17–26
    mobile vs. solid phase, 20
RXMA primers, 109

S

Scintillation counter
    DNA demethylase activity, 166
Second-dimension electrophoresis
    RLGS
        materials, 57
        methods, 63–64
Sephadex G50 columns
    preparation, 12
Sequence manipulation
    COBRA, 74–75
Sequencing
    bisulphite-assisted genomic
        sequencing, 150–152
    MSRF
        materials, 46
SmaI sites
    amplification, 101
Sodium bisulfite
    oxidizing, 79
    solution concentration, 79
    treatment
        COBRA, 78–80
Sodium metabisulfite, 79
Southern-blot hybridization, 143

Southern blotting
  MSRF
    materials, 45
Southwestern analysis
  DNA-binding proteins, 132
Southwestern blotting
  MeCp2-containing deacetylase
    materials, 134
    methods, 136
Southwestern oligo preparation
  MeCp2-containing deacetylase, 132–134
Superose 6 gel filtration
  Xenopus laevis extract fractionation, 138

T

Taq Gold, 109
TaqI
  restriction-enzyme site, 77
Test gel
  ligation PCR, 98
Thin-layer chromatography (TLC)
  developing tanks, 13
  nearest-neighbor analysis, 12–15
  nucleotide-3'-monophosphates position, 15
  plate stacking, 14

Transiently transfected HEK cells
  recombinant DNA demethylase purification, 173–175
Trichostatin A, 131
Tumor amplicon
  ovary, 89
Tumor suppressor genes
  inactivation, 43

U

Unmethylated cytosine residues
  incomplete conversion, 79
UV absorbance detector, 23

V–Z

Vertebrate genomes
  methylation, 111
Volatilization assay
  DNA demethylase activity, 165–166
Washing
  COBRA
    methods, 84–85
Xenopus laevis
  MeCP2-containing deacetylase, 131–140
  oocyte extract fractionation, 137–140
  oocyte extract preparation
    materials, 135
    methods, 137
Xist RNA, 3–4
Zetabind Membrane, 81